SpringerBriefs in Mathematics

SpringerBriefs in Mathematics showcases expositions in all areas of mathematics and applied mathematics. Manuscripts presenting new results or a single new result in a classical field, new field, or an emerging topic, applications, or bridges between new results and already published works, are encouraged. The series is intended for mathematicians and applied mathematicians.

More information about this series at http://www.springer.com/series/10030

Akbar Ali • Gary Chartrand • Ping Zhang

Irregularity in Graphs

 Springer

Akbar Ali
Department of Mathematics
University of Ha'il
Ha'il, Saudi Arabia

Gary Chartrand
Department of Mathematics
Western Michigan University
Kalamazoo, MI, USA

Ping Zhang
Department of Mathematics
Western Michigan University
Kalamazoo, MI, USA

ISSN 2191-8198   ISSN 2191-8201   (electronic)
SpringerBriefs in Mathematics
ISBN 978-3-030-67992-7   ISBN 978-3-030-67993-4   (eBook)
https://doi.org/10.1007/978-3-030-67993-4

Mathematics Subject Classification: 05C05, 05C10, 05C15, 05C45, 05C70, 05C78, 05C90

This Springer imprint is published by the registered company Springer Nature Switzerland AG
The registered company address is: Gewerbestrasse 11, 6330 Cham, Switzerland

Preface

What is often considered the first strictly theoretical paper dealing with graphs was written by the Danish mathematician Julius Petersen in 1891. This article was titled *Die Theorie der regulären Graphen* (the theory of regular graphs). In the 130 years since then, regular graphs have been a common and popular area of study. While regular graphs are typically considered to be graphs all of whose vertices have the same degree, a more general interpretation is that of graphs possessing some common characteristic throughout their structure. During the past several decades however, there has been increased interest in investigating graphs possessing a property that is, in a sense, opposite to regularity. It is this topic with which this book deals, giving rise to a study of what might be called irregularity in graphs.

The first paper to seriously investigate and call special attention to graphs that are opposite to regular graphs was written by Behzád and Chartrand in 1966. It was observed (as had been observed earlier) that the degrees of the vertices of a nontrivial graph cannot all be distinct, that is, there is no "irregular" graph. It was also observed that there are graphs in which exactly two vertices have the same degree, that is, there are "antiregular" graphs. These graphs together with a discussion of regular and irregular graphs are the topics of Chap. 1.

Even though no nontrivial graph exists whose vertices have distinct degrees, there are graphs in which the degrees of every two neighbors of each vertex are distinct. These are the highly irregular graphs. The existence of such graphs brings up the topic of locally irregular graphs. Despite the fact that the number of neighbors of all vertices of a nontrivial graph G cannot be distinct, it is possible for the subgraphs induced by the neighbors of all vertices of G to be distinct (or the same). These are the link-irregular and link-regular graphs. These graphs and the highly irregular graphs are the subjects of Chap. 2.

Because there are no nontrivial graphs whose vertices have distinct degrees, the number of subgraphs isomorphic to K_2 containing a specific vertex of a nontrivial graph G cannot be distinct among all vertices of G. This observation has led to the problem of investigating these numbers when the graph K_2 is replaced by other

v

graphs F, which in turn has led to the F-degree of a vertex in a graph and the study of F-regular graphs and the existence of F-irregular graphs. This is the topic of Chap. 3.

Despite the fact that there are no nontrivial irregular graphs, there are numerous irregular multigraphs (whose vertices have distinct degrees). In fact, for every connected graph G of order 3 or more, there exists an irregular multigraph with underlying graph G. Every multigraph M can be looked at as a weighted graph where for every two vertices u and v joined by one or more edges in M, the edge uv in the underlying graph G of M can be assigned a weight equal to the number of edges joining u and v in M. For a connected graph G of order 3 or more, this brought up the problem of determining the minimum positive integer k such that the edges of G can be assigned weights from the set $[k] = \{1, 2, \ldots, k\}$ so that the resulting weighted graph (equivalently, multigraph) is irregular. This concept, the irregularity strength of a graph, is the topic of Chap. 4.

In Chap. 4, each edge of a graph G is assigned a label (a weight) resulting in a weighted graph, and the degree of each vertex v of G in this weighted graph is obtained by summing the weights of the edges of G incident with v. The primary emphasis of Chap. 4 is to produce an irregular weighted graph by minimizing the weights used for the edges. There is no concern as to how large the degrees of the vertices of the resulting weighted graph may become provided the degrees of the vertices are distinct. In Chap. 5, the emphasis changes from minimizing the weights (referred to as colors here) used for the edges of a graph to minimizing the resulting degrees (colors) of the vertices of the graph but with the continued requirement that all vertex colors must be integers. This is accomplished by assigning positive integer colors to the edges of a graph to produce integer vertex colors which are the averages of the colors of the incident edges. The goal of Chap. 5 is then to minimize the largest vertex color under the condition that the vertex coloring is irregular, that is, the vertices have distinct colors. This results in the topic of rainbow mean colorings.

In Chap. 6, the emphasis changes once again. Beginning with assigning positive integer labels, weights, or colors to the edges of a graph and producing a vertex degree or color by adding or averaging the weights or colors assigned to the incident edges of a vertex, a vertex coloring is obtained by means of sets, namely a vertex color is the set of colors of the edges incident with the vertex. Once again, the main emphasis is to obtain an irregular vertex coloring by this means. This takes us to the topic of majestic colorings of graphs. Because, in a majestic coloring of a graph G, each edge of G is assigned a positive integer color and the color of a vertex is essentially obtained by a union of colors, this suggests the topic of assigning each edge of G a nonempty subset of positive integers as its color and obtaining a vertex coloring by taking the union of the sets of integers assigned to the edges incident with each vertex. This takes us to the topic of royal colorings of graphs, another primary topic of Chap. 6.

For the final two chapters of the book, there is once again a major shift in the study of irregularity in graphs by describing what might be considered the opposite of regular characteristics of a graph. In Chap. 7, the emphasis is on traversability in graphs. The first theoretical concept that arose in graph theory was essentially that of Eulerian graphs, which grew out of Leonhard Euler's 1736 solution of the famous Königsberg Bridge Problem. If a connected graph G is Eulerian, then there is always a circuit in which every edge of G is encountered exactly the same number of times (namely once), and only Eulerian graphs have this property. This brings up the question as to whether a graph G has a closed walk in which no two edges of G are encountered the same number of times and what the length of such a walk might be. This is the topic of Eulerian irregularity. A related topic is that of whether a graph G can contain a walk that encounters the vertices of G an unequal number of times. This leads us to the topic of Hamiltonian irregularity, the irregular version of graphs that contains a Hamiltonian path. This is the second primary topic of Chap. 7.

A popular area of study over a period of many years is that of whether certain graphs (often complete graphs) can be decomposed into subgraphs possessing some prescribed property. The most common of these decomposition problems deals with whether a certain graph G can be decomposed into a collection of subgraphs, each isomorphic to a given graph H. Such H-decomposable graphs deal with the topic of isomorphic decompositions, that is, all subgraphs in the decomposition are the same. This brings up the irregular version of this concept, namely irregular decompositions of graphs. An important related topic here is that of ascending subgraph decompositions and whether a graph G of size $\binom{n+1}{2}$ for some positive integer n can be decomposed into n subgraphs $G_1, G_2, \ldots, G_n$ of G, where G_i has size i for $1 \leq i \leq n$ and G_i is isomorphic to a subgraph of G_{i+1} for $1 \leq i \leq n - 1$. It is a long-standing conjecture that every graph of size $\binom{n+1}{2}$ possesses such a decomposition. This is a major topic of Chap. 8. This final chapter closes with a discussion of how this subject also has a connection with Ramsey numbers of graphs. For a graph G, the problem of determining, for every red-blue coloring of G, the largest positive integer k for which G has an ascending subgraph decomposition into k monochromatic subgraphs is introduced. Consequently, this topic deals with the question of whether a given graph possesses an irregular decomposition satisfying a Ramsey-type coloring property.

In addition to the many concepts touched upon throughout the book, the reader is encouraged to investigate conjectures and problems in greater depth.

Ha'il, Saudi Arabia   Akbar Ali

Kalamazoo, MI, USA   Gary Chartrand

Kalamazoo, MI, USA   Ping Zhang
November 17, 2020

Contents

Chapter 1
Introduction

1.1 Prologue

Among the most popular topics in graph theory are those dealing with concepts based on "all things the same". Some of the best known examples of these topics concern those graphs all of whose vertices have the same degree, graphs that can be decomposed into subgraphs that are all the same (isomorphic), and graphs whose edges are colored with a specific number of colors that guarantee the existence of a particular subgraph all of whose edges are colored the same.

To find the theoretical origin of such studies, one must go back many years—indeed to the late nineteenth century [17]. It is generally understood that the first paper dealing with the theory of graphs (that is, of a theoretical nature on graphs) was *Die Theorie der regulären Graphen* (the theory of regular graphs) written by the Danish mathematician Julius Petersen and appearing in *Acta Mathematica* in 1891. This paper is significant for a number of reasons. According to the description in [14], Petersen was working on a problem in invariant theory. In 1889 he met James Joseph Sylvester, who, some 11 years earlier, was using graphs to analyze invariant theory. In fact, it was at that time that Sylvester introduced the term *graph*. This eventually led to Petersen's famous paper. Although written in German, Petersen wrote the word *graph* in English.

The topic that interested Petersen in his paper was *regular graphs* (graphs whose vertices have the same degree) that contain a certain type of subgraph or can be decomposed into certain isomorphic subgraphs. His most famous theorem deals with the fact that certain cubic (3-regular) graphs contain a 1-factor (a 1-regular spanning subgraph). A *bridge* in a connected graph is an edge whose removal results in a disconnected graph.

Theorem 1.1 (Petersen) *Every cubic graph with at most two bridges contains a 1-factor.*

A. Ali et al., *Irregularity in Graphs*, SpringerBriefs in Mathematics,
https://doi.org/10.1007/978-3-030-67993-4_1

Fig. 1.1 Two drawings of the
Petersen graph

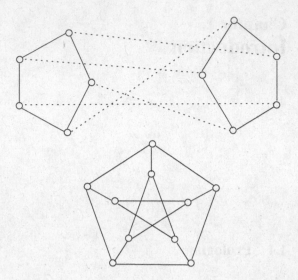

It is easy to see that if a cubic graph has three bridges, then it need not have a 1-factor. Petersen also encountered cubic graphs in an 1898 paper [18] when he gave an example of a cubic graph without bridges that is not 1-factorable (decomposable into three 1-factors), thereby providing a counterexample to a "theorem" by Peter Tait [21]. Tait believed that every cubic graph is 1-factorable, but Petersen's graph (the Petersen graph) showed that this is not true. Figure 1.1 shows the drawing of the Petersen graph given by Petersen, under which is the usual drawing of this famous graph. Even though the graph is named for Petersen, this graph appeared in the literature in 1886 in a paper by Kempe [9], famous himself for an attempted solution of the Four Color Problem.

Petersen [17] also proved another theorem dealing with regular graphs and factorizations. A graph G is 2-*factorable* if it can be decomposed into 2-factors (2-regular spanning subgraphs) of G.

Theorem 1.2 (Petersen) *If G is an r-regular graph where $r \geq 2$ is even, then G is 2-factorable.*

While the graph theory literature has numerous papers dealing with graphs that are regular, graphs having decompositions into isomorphic subgraphs, and related topics, the past half-century has shown an increased interest in topics that are, in a variety of ways, opposite to this. It is this area, namely irregularity in graphs, that is our primary topic here. First, we provide a foundation for this topic.

1.2 Degrees and Degree Sets

There are many numbers associated with each graph. The most common are the number of vertices (its order), the number of edges (its size), and the degrees of its vertices. The degree of a vertex v in a graph G (the number of edges incident with v) is denoted by $\deg_G v$, or $\deg v$ if the graph G under consideration is clear. These numbers are all represented in a basic observation in graph theory.

Proposition 1.1 *Let G be a graph of order n and size m. If the degrees of all n vertices are added, the resulting sum is $2m$.*

A vertex in a graph is even if its degree is even and is odd otherwise. An immediate consequence of Proposition 1.1 is the following fact.

Corollary 1.1 *Every graph contains an even number of odd vertices.*

While in a graph of order n, the degree of a vertex must be some number in the set $[0, n-1] = \{0, 1, \ldots, n-1\}$, there is no other restriction on what the degree of a vertex in a graph of order n can be. The *degree set* $\mathscr{D}(G)$ of a graph G is the set of degrees of the vertices of G. Therefore, for every graph G, it follows that $\emptyset \neq \mathscr{D}(G) \subseteq [0, n-1]$. If G has no isolated vertices (vertices of degree 0), then $\emptyset \neq \mathscr{D}(G) \subseteq [n-1] = [1, n-1]$. This brings up the question:

> For which finite sets S of positive integers whose largest element is $n-1$, does there exist a graph G of order n such that $\mathscr{D}(G) = S$?

This question was answered in [8]. For two vertex-disjoint graphs F and H, the *join* of F and H (the graphs F and H and all edges joining a vertex of F and a vertex of H) is denoted by $F \vee H$ and the *union* of F and H (consisting only of F and H) is denoted by $F + H$. For a graph G, its complement is denoted by $\overline{G}$.

Theorem 1.3 *For every set $S = \{a_1, a_2, \ldots, a_k\}$, $k \geq 1$, of positive integers with $a_1 < a_2 < \cdots < a_k$, there exists a graph G of order $a_k + 1$ with $\mathscr{D}(G) = S$.*

Proof We verify this by induction. For $k = 1$, the complete graph K_{a_1+1} has this property; while for $k = 2$, the graph $K_{a_1} \vee \overline{K}_{a_2-a_1+1}$ has the desired property. Assume, for an integer $k \geq 2$ and for every set S^* consisting of i positive integers with $1 \leq i \leq k$, where a_i is the largest element of S^*, that there is a graph F of order $a_i + 1$ such that $\mathscr{D}(F) = S^*$. Now let $S = \{b_1, b_2, \ldots, b_{k+1}\}$ be a set of $k + 1$ positive integers with $b_1 < b_2 < \cdots < b_{k+1}$. Then $S' = \{b_2 - b_1, b_3 - b_1, \ldots, b_{k+1} - b_1\}$ is a set of k positive integers whose largest element is $b_{k+1} - b_1$. By the induction hypothesis, there is a graph H of order $b_{k+1} - b_1 + 1$ such that $\mathscr{D}(H) = S'$. Hence, the graph $G = K_{b_1} \vee (\overline{K}_{b_{k+1}-b_k} + H)$ has order $b_{k+1} + 1$ and $\mathscr{D}(G) = S$. $\qquad\square$

1.3   Regular and Irregular Graphs

By Theorem 1.3, not only is every finite nonempty set S of positive integers the degree set of some graph, but there is a graph with degree set S whose order is one more than the largest element of S. Graphs whose degree set consists of a single integer have always been of interest to researchers. These are the *regular graphs*. If the single integer is r, then the graphs are *r-regular graphs*. The 0-regular graphs consist only of isolated vertices. Each component of a 1-regular graph is the complete graph K_2 of order 2; while each component of a 2-regular graph is a cycle. A *k-regular spanning subgraph*, $k \geq 1$, of a graph G is a *k-factor* of G. If a graph can be decomposed into k-factors, then it is *k-factorable*. The 3-regular or cubic graphs are considerably more complicated and more interesting. Two well-known facts concerning cubic graphs, obtained in [17] and [21], respectively, are the following, the first of which is a consequence of Theorem 1.1.

Theorem 1.4 *Every bridgeless cubic graph contains a 1-factor.*

A *proper coloring* of the vertices of a graph G is one in which every two adjacent vertices of G are colored differently. A graph G is *4-colorable* if there is a proper coloring of the vertices of G with four or fewer colors. A *proper coloring* of the edges of a graph is one in which every two adjacent edges are colored differently.

Theorem 1.5 *Every planar graph is 4-colorable if and only if the edges of every bridgeless cubic planar graph has a proper 3-coloring.*

Of course, every complete graph K_n of order n is $(n - 1)$-regular. If n is even, then K_n is 1-factorable; while if $n \geq 3$ is odd, then K_n is not only 2-factorable, K_n can be decomposed into Hamiltonian cycles (spanning cycles). From this, we have the following observation.

Proposition 1.2 *For every two integers r and n with $0 \leq r < n$, there exists an r-regular graph of order n if and only if rn is even.*

While regular graphs have been shown to be important, interesting, and a subject of study for over a century, those graphs that are, in a sense, opposite to regular graphs have gained interest in recent decades. It is graph theory folklore that for every integer $n \geq 2$, there is no graph of order n all of whose vertices have distinct degrees. These non-existent graphs were first looked at more formally in [4], where these graphs were called *perfect* (so that one could say *no graph is perfect*). Later, however, when it became clear that the term "perfect", as used by Berge [5], had been accepted to have an entirely different meaning, the perfect graphs in [4] became known as *irregular graphs*.

Theorem 1.6 *For every integer $n \geq 2$, there is no irregular graph of order n.*

Proof Assume, for some integer $n \geq 2$, that there exists an irregular graph G of order $n \geq 2$. Then the degree set of the n vertices of G must be $\{0, 1, \ldots, n - 1\}$, implying that G has both an isolated vertex and a vertex of degree $n - 1$, which is impossible.   □

In 1988, the British mathematics educator David Wells listed 24 theorems in an article [22], asking the readers to vote on which of these was the most beautiful. In 1990, he wrote a follow-up article [23] giving the results of this survey. The theorem that was voted #1 was $e^{i\pi} + 1 = 0$. This theorem by Leonhard Euler involves the five most famous numbers in mathematics, namely $0, 1, \pi, e, i$. Another theorem listed among the 24 theorems was the following:

> At any party, there is a pair of people who have the same number of friends present.

This, of course, is equivalent to the result stated in Theorem 1.6.

While we have seen that no graph exists in which all vertices have distinct degrees, many related questions arise. It is the goal of this book to discuss some of these questions as well as topics and concepts suggested by them.

1.4 Antiregular Graphs

In the preceding section, we saw in Theorem 1.6 that there is no graph of order 2 or more whose vertices have distinct degrees, that is, no graph is irregular. Equivalently, for each integer $n \geq 2$, there is no graph G of order n whose degree set satisfies $|\mathscr{D}(G)| = n$. On the other hand, for each integer $n \geq 2$, there is a graph G of order n such that $|\mathscr{D}(G)| = n - 1$. In fact, if a graph G has this property, then its complement $\overline{G}$ satisfies $|\mathscr{D}(\overline{G})| = n - 1$ as well. These graphs have been referred to by many names in the literature. Here, we refer to these graphs as antiregular graphs. Formally then, a graph G of order $n \geq 2$ is *antiregular* if $|\mathscr{D}(G)| = n - 1$. That is, in an antiregular graph there are exactly two vertices having the same degree.

It is easy to give examples of antiregular graphs. There are only two graphs of order $n = 2$, namely K_2 and $\overline{K}_2$, and both are antiregular; while for $n = 3$, the graphs P_3 and $\overline{P}_3 = K_1 + K_2$ are antiregular. In fact, these four graphs are the only antiregular graphs of order 2 or 3. Indeed, for each integer $n \geq 2$, there are exactly two non-isomorphic antiregular graphs of order n, namely a connected antiregular graph G_n and its disconnected complement $\overline{G}_n$. The eight antiregular graphs of order n where $2 \leq n \leq 5$ and the connected antiregular graph G_6 of order 6 are shown in Fig. 1.2. Observe, for example, that $G_6 = \overline{G}_5 \vee K_1$ is the join of $\overline{G}_5$ and K_1.

As we mentioned, for each integer $n \geq 2$, there are exactly two antiregular graphs of order n.

Theorem 1.7 *For every integer $n \geq 2$, there are exactly two non-isomorphic antiregular graphs of order n, namely one connected and one disconnected.*

Proof The statement is obvious for small values of n, say $2 \leq n \leq 5$. Assume, however, that the statement is false. Then there is a smallest integer $n \geq 6$ for

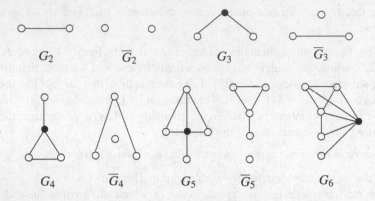

Fig. 1.2 Some antiregular graphs

which there exist two non-isomorphic connected antiregular graphs F_n and H_n of order n. Since F_n and H_n are not isomorphic, their complements $\overline{F}_n$ and $\overline{H}_n$ are also not isomorphic. Because each of F_n and H_n contains a vertex of degree $n-1$, each of $\overline{F}_n$ and $\overline{H}_n$ contains an isolated vertex and one other component, which, necessarily, is the unique connected antiregular graph G_{n-1} of order $n-1$ in each case. Thus, $\overline{F}_n$ and $\overline{H}_n$ are isomorphic, which produces a contradiction.   □

The proof of Theorem 1.7 tells us how the connected antiregular graph G_{n+1} of order $n+1$ can be obtained from the connected antiregular graph G_n of order n for each integer $n \geq 2$, namely, G_{n+1} can be obtained from G_n by adding a new vertex to $\overline{G}_n$ and joining this vertex to every vertex of $\overline{G}_n$. Thus, $G_{n+1} = \overline{G}_n \vee K_1$ where $\overline{G}_n = G_{n-1} + K_1$. With this information, we can now recursively construct all connected antiregular graphs G_n, $n \geq 2$, and their complements. For even integers n, we begin with $G_2 = K_2$. Adding an isolated vertex to G_2 gives us $\overline{G}_3 = G_2 + K_1$. Adding a vertex to $\overline{G}_3$ and joining this vertex to every vertex of $\overline{G}_3$ gives us $G_4 = \overline{G}_3 \vee K_1$. By adding an isolated vertex to G_4, we obtain $\overline{G}_5 = G_4 + K_1$. Continuing this procedure, we obtain the sequence $G_2, \overline{G}_3, G_4, \overline{G}_5, G_6, \ldots$. For odd integers n, we begin with $G_3 = P_3$ and proceeding as above gives us the sequence $G_3, \overline{G}_4, G_5, \overline{G}_6, G_7, \ldots$.

In Sect. 1.3, we described a theorem stated in terms of a party. There is a related party problem:

> At a party attended by n people, every person has at least one friend at the party. One person at the party, Pat, asks the other $n-1$ people: How many people at this party are friends of yours? All $n-1$ responses are different. How many people are friends of Pat?

We can also look at this question in terms of graph theory. Let G_n be a graph of order n whose vertices are the people at the party. Two vertices are joined by an edge if these two people are friends. Since $n-1$ people at the party have different numbers of friends at the party, the degrees of these $n-1$ vertices are different and so G_n is antiregular. Because every person has at least one friend at the party, G_n

has no isolated vertices. Therefore, G_n is the unique connected antiregular graph of order n and the degree of the vertex Pat is the same as the degree of some other vertex of G_n. The question then is: What is the repeated degree of two vertices in the connected antiregular graph of order n?

Theorem 1.8 *Let $n \geq 2$ be an integer. The connected antiregular graph G_n of order n has two vertices of degree $\left\lfloor \dfrac{n}{2} \right\rfloor$.*

Proof We proceed by induction on $n \geq 2$. This is clear for small values of n, say $n = 2, 3, 4, 5$ (see Fig. 1.2). Assume for an integer $n \geq 5$ that G_n has two vertices u and v of degree $\lfloor n/2 \rfloor$. Since $G_{n+1} = \overline{G}_n \vee K_1$, the degree of u and v in G_{n+1} is

$$\left(n - 1 - \left\lfloor \frac{n}{2} \right\rfloor\right) + 1 = n - \left\lfloor \frac{n}{2} \right\rfloor = \left\lceil \frac{n}{2} \right\rceil = \left\lfloor \frac{n+1}{2} \right\rfloor,$$

as desired.   □

We now describe some other properties possessed by the connected antiregular graph G_n of order n. The *clique number* $\omega(G)$ of a graph G is the maximum order of a complete subgraph (clique) of G. A set S of vertices in a graph G is an *independent set* if no two vertices in S are adjacent. The *independence number* $\alpha(G)$ of G is the maximum cardinality of an independent set of vertices in G. First, we consider the clique number and the independence number of the connected antiregular graph G_n of order n. For small values of n, it is readily seen that $\omega(G_n) = \lceil (n + 1)/2 \rceil$ and $\alpha(G_n) = \lfloor (n + 1)/2 \rfloor$. We now show that this is true in general.

Theorem 1.9 *For every integer $n \geq 2$,*

$$\omega(G_n) = \left\lceil \frac{n+1}{2} \right\rceil \quad and \quad \alpha(G_n) = \left\lfloor \frac{n+1}{2} \right\rfloor.$$

Consequently, $\omega(G_n) + \alpha(G_n) = n + 1$.

Proof We proceed by induction on $n \geq 2$. As mentioned before, $\omega(G_n) = \left\lceil \dfrac{n+1}{2} \right\rceil$ and $\alpha(G_n) = \left\lfloor \dfrac{n+1}{2} \right\rfloor$ for small values of n, say $2 \leq n \leq 5$ (see Fig. 1.2). Assume for some integer $n \geq 5$ that

$$\omega(G_k) = \left\lceil \frac{k+1}{2} \right\rceil \quad and \quad \alpha(G_k) = \left\lfloor \frac{k+1}{2} \right\rfloor$$

for all integers k with $2 \leq k \leq n$. We now consider the graph G_{n+1}, where, as we have seen, $G_{n+1} = \overline{G}_n \vee K_1$. Since $\overline{G}_n = G_{n-1} + K_1$, it follows that

$$\omega(\overline{G}_n) = \left\lceil \frac{n}{2} \right\rceil \quad and \quad \alpha(\overline{G}_n) = \left\lfloor \frac{n}{2} \right\rfloor + 1 = \left\lfloor \frac{n+2}{2} \right\rfloor.$$

Consequently,

$$\omega(G_{n+1}) = \left\lceil \frac{n}{2} \right\rceil + 1 = \left\lceil \frac{n+2}{2} \right\rceil \text{ and } \alpha(G_{n+1}) = \left\lfloor \frac{n+2}{2} \right\rfloor,$$

producing the desired result.   □

The *chromatic number* $\chi(G)$ of a graph G is the minimum number of colors in a proper coloring of the vertices of G. Thus, $\chi(G) \geq \omega(G)$. Since $\omega(G_n) = \lceil (n+1)/2 \rceil$ for the graph G_n of order $n \geq 2$, it follows that $\chi(G_n)$ is at least $\lceil (n+1)/2 \rceil$. However, there is equality here for each such integer n.

Theorem 1.10 *For every integer* $n \geq 2$,

$$\chi(G_n) = \left\lceil \frac{n+1}{2} \right\rceil.$$

Proof Here too, it is easy to see that $\chi(G_n) = \left\lceil \dfrac{n+1}{2} \right\rceil$ for small values of n, say $2 \leq n \leq 5$ (see Fig. 1.2). Once again, we proceed by induction on n. Assume for some integer $n \geq 5$ that $\chi(G_k) = \left\lceil \dfrac{k+1}{2} \right\rceil$ for all integers k with $2 \leq k \leq n$. We now consider the graph G_{n+1}, where, as we have seen, $G_{n+1} = \overline{G}_n \vee K_1$. Since $\overline{G}_n = G_{n-1} + K_1$, $\chi(G_{n-1}) = \left\lceil \dfrac{n}{2} \right\rceil$ and the isolated vertex of $\overline{G}_n$ can be added to any color class of G_{n-1}, we have $\chi(\overline{G}_n) = \left\lceil \dfrac{n}{2} \right\rceil$. Since the vertex added to $\overline{G}_n$ in the construction of G_{n+1} that is joined to every vertex of $\overline{G}_n$ must belong to a color class of its own, it follows that $\chi(G_{n+1}) = \left\lceil \dfrac{n}{2} \right\rceil + 1 = \left\lceil \dfrac{n+2}{2} \right\rceil$.   □

A graph G is *perfect* if $\chi(H) = \omega(H)$ for every induced subgraph H of G. (This is the definition of a perfect graph, as defined by Berge.) It was proved by Lovász [11] that a graph G is perfect if and only if $\overline{G}$ is perfect. For every induced subgraph H of G_n where $2 \leq n \leq 5$ (see Fig. 1.2), one can see that $\chi(H) = \omega(H)$, that is, G_n is perfect for $2 \leq n \leq 5$. In fact, this is the case for every graph G_n where $n \geq 2$.

Theorem 1.11 *For every integer* $n \geq 2$, *the graph* G_n *is perfect*.

Proof We proceed by induction on $n \geq 2$. We have already noted that G_n is perfect for $2 \leq n \leq 5$. Assume that G_n is perfect for an integer $n \geq 5$. We show that G_{n+1} is perfect. Let H be an induced subgraph of G_{n+1}. Once again, recall that $G_{n+1} = \overline{G}_n \vee K_1$. First, suppose that H is also an induced subgraph of $\overline{G}_n$. Since G_n is perfect by the induction hypothesis, it follows that $\overline{G}_n$ is perfect and so are both $\overline{H}$ and H. Thus, $\chi(H) = \omega(H)$. Next, suppose that H contains the vertex $v \in V(G_{n+1}) - V(\overline{G}_n)$. Then $F = H - v$ is an induced subgraph of $\overline{G}_n$ and so $\chi(F) =$

$\omega(F)$. Since $\chi(H) = \chi(F) + 1$ and $\omega(H) = \omega(F) + 1$, it follows that $\chi(H) = \omega(H)$ and so G_{n+1} is perfect.   □

It is also readily observed that if H_k is an induced subgraph of order k in G_n (or in $\overline{G}_n$) where $2 \leq n \leq 5$ and $1 \leq k \leq n$, then H_k contains either an isolated vertex or a vertex of degree $k - 1$. This too is true in general.

Theorem 1.12 *Let k and $n \geq 2$ be integers with $1 \leq k \leq n$. If H_k is an induced subgraph of order k in G_n for any integer k with $1 \leq k \leq n$, then H_k contains either an isolated vertex or a vertex of degree $k - 1$.*

Proof We proceed by induction on $n \geq 2$. We have already noted that the statement is true for all integers n with $2 \leq n \leq 5$. Assume for an integer $n \geq 5$ that if H_k is an induced subgraph of order k in G_n for any integer k with $1 \leq k \leq n$, then H_k contains either an isolated vertex or a vertex of degree $k - 1$. We now consider the graph $G_{n+1} = \overline{G}_n \vee K_1$. Let F_k be an induced subgraph of order k in G_{n+1}. If $k = 1$, then $F_k = K_1$ consists of a single isolated vertex. Hence, we may assume that $2 \leq k \leq n$. Suppose first that $F_k \subseteq \overline{G}_n$. Since every induced subgraph of order k in G_n contains either an isolated vertex x or a vertex y of degree $k - 1$, it follows that in F_k, the vertex x has degree $k - 1$ or y is an isolated vertex. We may assume that F_k contains the vertex v of G_{n+1} that does not belong to $\overline{G}_n$. Since v is adjacent to every vertex of $\overline{G}_n$, it follows that v has degree $k - 1$ in F_k.   □

We now describe two different ways to construct the graphs G_n for $n \geq 3$. First, let G be a graph of order $n \geq 3$ with $V(G) = \{v_1, v_2, \ldots, v_n\}$ where $E(G) = \{v_i v_j : i + j \geq n + 1\}$. Then the edge set of the complement $\overline{G}$ is $E(\overline{G}) = \{v_i v_j : i + j \leq n\}$. The graphs $G = G_n$ and $\overline{G} = \overline{G}_n$ are precisely the two antiregular graphs of order n. For $n = 5$, for example, the graphs G_5 and $\overline{G}_5$ defined in this way are given in Fig. 1.3.

By defining the connected antiregular graph G_n, $n \geq 4$, in this manner, one might ask what the construction of the resulting graph would be if, in the definition of $E(G_n)$, we were to replace $i + j \geq n + 1$ by $i + j \geq n + 2$ or $i + j \geq n$, for example. In the first case, the resulting graph has a direct connection to antiregular graphs; while in the second case, the resulting graph has another interesting property. More generally, let n and k be nonnegative integers such that n is sufficiently large (in terms of k). Recall that if G is a graph of order n with $V(G) = \{v_1, v_2, \ldots, v_n\}$ and

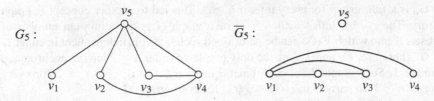

Fig. 1.3 The two antiregular graphs of order 5

$E(G) = \{v_i v_j : i + j \geq n + 1\}$, then $G = G_n$, the connected antiregular graph of order n.

- First, suppose that $E(G) = \{v_i v_j : i + j \geq n + k\}$, where $1 \leq k \leq n - 1$. Then $G = (k - 1)K_1 + G_{n-k+1}$.
- Next, suppose that $E(G) = \{v_i v_j : i + j \geq n - k\}$, where $0 \leq k \leq n - 3$. Then $G = K_{k+1} \vee G_{n-k-1}$. (If we define $K_0 = \emptyset$, then we could take $-1 \leq k \leq n-3$). If $0 \leq k \leq n - 3$, then two degrees are repeated, one degree is $n - 1$ and the other repeated as a pair.

To describe a second way to construct the graphs G_n, we define a graph G of order n for each integer $n \geq 3$.

- First, assume that $n \geq 4$ is even. Then $n = 2k$ where $k \geq 2$. Let $V(G) = A \cup B$ where $A = \{u_1, u_2, \ldots, u_{k+1}\}$ and $B = \{v_1, v_2, \ldots, v_{k-1}\}$ such that the subgraph of G induced by A is $G[A] = K_{k+1}$ and the subgraph of G induced by B is $G[B] = \overline{K}_{k-1}$. In addition, $u_i v_j \in E(G)$ if and only if $i \leq j$. Then

$$\deg_G u_i = k + (k - i) = 2k - i \text{ for } 1 \leq i \leq k - 1$$

$$\deg_G u_k = \deg_G u_{k+1} = k$$

$$\deg_G v_j = j \text{ for } 1 \leq j \leq k - 1.$$

Thus, $G \cong G_n$. This is illustrated in Fig. 1.4a.
- Next, assume that $n \geq 3$ is odd. Then $n = 2k+1$ where $k \geq 1$. Let $V(G) = A \cup B$ where $A = \{u_1, u_2, \ldots, u_{k+1}\}$ and $B = \{v_1, v_2, \ldots, v_k\}$ where $G[A] = K_{k+1}$ and $G[B] = \overline{K}_k$. In addition, $u_i v_j \in E(G)$ if and only if $i \leq j$. Then

$$\deg_G u_i = k + (k + 1 - i) = 2k + 1 - i \text{ for } 1 \leq i \leq k + 1$$

$$\deg_G u_k = \deg_G u_{k+1} = k$$

$$\deg_G v_j = j \text{ for } 1 \leq j \leq k.$$

Thus, $G \cong G_n$. This is illustrated in Fig. 1.4b.

From the construction of the graph G_n, $n \geq 3$, we see that the vertex set $V(G_n)$ of G_n has a partition $\{A, B\}$ into two sets A and B, where A is independent in $\overline{G}_n$ and B is independent in G_n. A graph with this property is called a *split graph* and so G_n is a *split graph* for every integer $n \geq 3$. This led to another concept in graph theory. The *cochromatic number* $z(G)$ of a graph G is the minimum number of subsets A into which $V(G)$ can be partitioned such that A is independent in either G or $\overline{G}$. Therefore, K_n and $\overline{K}_n$ are the only graphs of order $n \geq 2$ having cochromatic number 1. Split graphs have cochromatic number 2. Thus, $z(G_n) = 2$ for every integer $n \geq 3$. In particular, $z(G) \leq \chi(G)$ for every graph G.

Equivalently, for a graph G, the cochromatic number $z(G) = k$ if k is the minimum number of subsets $V_1, V_2, \ldots, V_k$ into which $V(G)$ can be partitioned such that $\chi(G[V_i]) = 1$ or $\chi(\overline{G}[V_i]) = 1$ for $1 \leq i \leq k$. By requiring $\chi(G[V_i]) \leq 2$

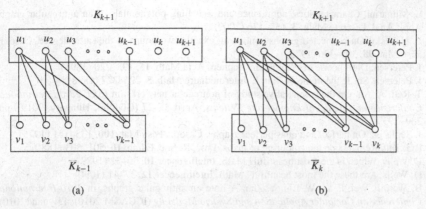

Fig. 1.4 Constructing the graph G_n of order n. (**a**) $n \geq 4$ is even. (**b**) $n \geq 3$ is odd

or $\chi(\overline{G}[V_i]) \leq 2$ for $1 \leq i \leq k$, we have the *bipartite cochromatic number* $z_2(G)$ of the graph G.

Additional information on antiregular graphs is also given in [1–4, 6, 10, 12, 13, 15, 16, 19, 20, 24].

References

1. A. Ali, A survey of antiregular graphs. Contrib. Math. **1**, 67–79 (2020)
2. C.O. Aguilar, M. Ficarra, N. Schurman, B. Sullivan, The role of the anti-regular graph in the spectral analysis of threshold graphs. Linear Algebra Appl. **588**, 210–223 (2020)
3. C.O. Aguilar, J. Lee, E. Piato, B.J. Schweitzer, Spectral characterizations of anti-regular graphs, Linear Algebra Appl. **557**, 84–104 (2018)
4. M. Behzad, G. Chartrand, No graph is perfect. Am. Math. Monthly **74**, 962–963 (1967)
5. C. Berge, Some classes of perfect graphs, in *Six Papers on Graph Theory* (Indian Statistical Institute, Calcutta, 1963), pp. 1–21
6. A. Berman, X.-D. Zhang, A note on degree antiregular graphs. Linear Multilinear Algebra **47**, 307–311 (2000)
7. G. Chartrand, P. Zhang, *Chromatic Graph Theory*, 2nd edn (Chapman & Hall/CRC Press, Boca Raton, 2020).
8. S.F. Kapoor, A.D. Polimeni, C.E. Wall, Degree sets for graphs. Fund. Math. **95**, 189–194 (1977)
9. A.B. Kempe, A memoir on the theory of mathematical form. Philos. Trans. R. Soc. Lond. **177**, 1–70 (1886)
10. V.E. Levit, E. Mandrescu, On the independence polynomial of an antiregular graph. Carpathian J. Math. **28**, 279–288 (2012)
11. L. Lovász, A characterization of perfect graphs. J. Combin. Theory Ser. B **13**, 95–98 (1972)
12. R. Merris, Doubly stochastic graph matrices II. Linear Multilinear Algebra **45**, 275–285 (1998)
13. R. Merris, Antiregular graphs are universal for trees, Univ. Beograd. Publ. Elektrotehn. Fak. Ser. Mat. **14**, 1–3 (2003)
14. H.M. Mulder, Julius Petersen's theory of regular graphs. Discrete Math. **100**, 157–175 (1992)

15. E. Munarini, Characteristics, admittance and matching polynomials of an antiregular graph. Appl. Anal. Discrete Math. **3**, 157–176 (2009)
16. L. Nebeský, On connected graphs containing exactly two points of the same degree. Časopis Pést. Mat. **98**, 305–306 (1973)
17. J. Petersen, Die Theorie der regulären Graphen. Acta Math. **15**, 193–220 (1891)
18. J. Petersen, Sur le théoréme de Tait. L' Intermédiaire Math. **5**, 225–227 (1898)
19. T. Réti, A. Ali, On the comparative study of nonregular networks, in *IEEE 23rd International Conference on Intelligent Engineering Systems*, April 25–27 (Gödöllő, Hungary, 2019), pp. 289–293
20. J. Sedláček, On perfect and quasiperfect graphs. Časopis Pést. Mat. **100**, 135–141 (1975)
21. P.G. Tait, Remarks on the colouring of maps. Proc. R. Soc. Edinb. **10**, 501–503 (1880)
22. D. Wells, Which is the most beautiful? Math. Intelligencer **10**, 30–31 (1988)
23. D. Wells, Are these the most beautiful? Math. Intelligencer **12**, 37–41 (1990)
24. B. Wen, F. Wei, F. Li, W. Liu, Y. Zhu, A note on antiregular graphs, in *2010 International Conference on Computer Application and System Modeling* (ICCASM 2010), Taiyuan (2010), pp. V11-420–V11-421

Chapter 2
Locally Irregular Graphs

We saw in Chap. 1 that it is impossible for the degrees of every two vertices of a nontrivial graph G to be different. However, rather than considering all vertices of G, if one were to consider the vertices individually and investigate the degrees of the neighbors or the structure of the subgraph induced by the neighbors of a vertex, an entirely different outcome is possible. These are the topics of the current chapter.

2.1 Highly Irregular Graphs

One obvious property of a regular graph, say an r-regular graph G with $r \geq 2$, is that all neighbors of each vertex of G have the same degree, namely r. Of course, if G is a graph with the property that the degree of each neighbor of every vertex of G is r, then G is r-regular. Another possibility is that the degrees of the neighbors of every vertex of a nontrivial connected graph G are the same, but these degrees are different for some pair of vertices of G. For example, let u and v be two adjacent vertices of the complete bipartite graph $K_{s,t}$ where $s \neq t$. We may assume that $\deg u = s$ and $\deg v = t$. Thus, all vertices in $N(u)$ have degree t and all vertices in $N(v)$ have degree s. In fact, this situation can only occur for a connected (not necessarily complete) bipartite graph with partite sets U and W where every vertex of U has degree s, say, and every vertex of W has degree t such that $s|U| = t|W|$. The graph of Fig. 2.1 has this property where $\{s, t\} = \{2, 3\}$.

While there is no nontrivial graph whose vertices have distinct degrees, it is possible for the degrees of the neighbors of every vertex of a graph to be distinct. That is, while no graph is (globally) irregular, a graph can be locally irregular. A graph with this property is called highly irregular. Specifically, a graph G is *highly irregular* if for every vertex v of G, no two neighbors of v have the same degree. Two highly irregular graphs are shown in Fig. 2.2.

A. Ali et al., *Irregularity in Graphs*, SpringerBriefs in Mathematics,
https://doi.org/10.1007/978-3-030-67993-4_2

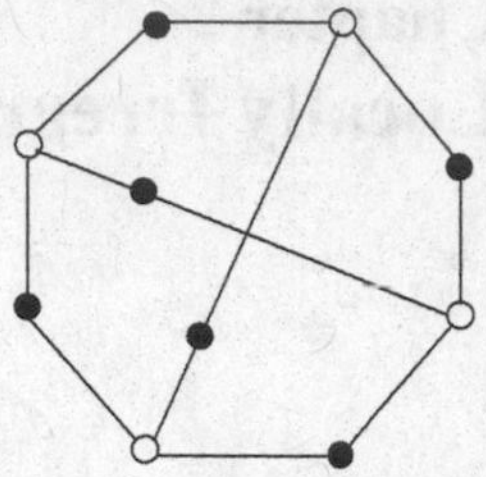

Fig. 2.1 A bipartite graph in
which the neighbors of each
vertex have the same degree

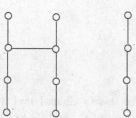

Fig. 2.2 Two highly irregular
graphs

The concept of highly irregular graphs was introduced and first studied in [2]. In
this case, the terminology actually preceded the concept. Yousef Alavi, a co-author
of [2], had the habit of referring to many situations as "highly irregular" and another
co-author, Ronald Graham, felt that it would be of interest to introduce and study
the class of "highly irregular" graphs. All that was needed was the definition. Highly
irregular graphs arose from this. First, we present a number of preliminary results
about highly irregular graphs, which appeared in [2].

Observation 2.1 *A vertex of maximum degree Δ in a highly irregular graph is
adjacent to exactly one vertex of each of the degrees $1, 2, \ldots, \Delta$.*

Proposition 2.1 *If G is a highly irregular graph of order n and maximum
degree $\Delta \geq 1$, then G contains at least two vertices of each of the
degrees $1, 2, \ldots, \Delta$ and consequently $n \geq 2\Delta$.*

Proof Since P_2 is the only highly irregular connected graph of order 2 having
maximum degree 1, we may assume that G is a highly irregular graph of order n
and maximum degree $\Delta \geq 2$. It follows by Observation 2.1 that G contains two
adjacent vertices u and v of degree Δ. Suppose that $N(u) = \{u_1, u_2, \ldots, u_{\Delta-1}, v\}$
and $N(v) = \{v_1, v_2, \ldots, v_{\Delta-1}, u\}$, where $\deg u_i = \deg v_i = i$ for $1 \leq i \leq \Delta - 1$.
Since G is highly irregular, no vertex of G is adjacent to two vertices of degree Δ.
This implies that

$$\{u_1, u_2, \ldots, u_{\Delta-1}\} \cap \{v_1, v_2, \ldots, v_{\Delta-1}\} = \emptyset.$$

Therefore, G contains at least two vertices of each of the degrees $1, 2, \ldots, \Delta$ and
so $n \geq 2\Delta$.   □

A fundamental question here is that of determining the existence of highly irregular graphs with some prescribed property. For example, for every positive integer Δ, there is a highly irregular graph G with $\Delta(G) = \Delta$.

Proposition 2.2 *For each positive integer Δ, there is a highly irregular graph with maximum degree Δ.*

Proof We proceed by induction on Δ. For $\Delta = 1$, we saw that P_2 has the desired property. Suppose, for some positive integer Δ, that there is a highly irregular graph H with maximum degree Δ. We show that there is a highly irregular graph G with maximum degree $\Delta + 1$. Let H' be another copy of H. The graph G is constructed from H and H' by joining a vertex u of degree Δ in H with its corresponding vertex u' in H'. Thus, the maximum degree of G is $\Delta + 1$. It remains to show that G is highly irregular. Let $v \in V(G)$, say $v \in V(H)$, and let x and y be two neighbors of v in G. Then at most one of x and y belongs to H'. First, suppose that $x, y \in V(H)$. Since H is highly irregular, we may assume that $\deg_H x < \deg_H y$. Then (i) $\deg_G x = \deg_H x$ and (ii) $\deg_G y = \deg_H y$ if $y \neq u$ or $\deg_G y = \deg_H y + 1$ if $y = u$. Hence, $\deg_G x < \deg_G y$. Next, suppose that one of x and y belongs to H', say $y \in V(H')$. Then $v = u$, $y = u'$, and so $\deg_G x = \deg_H x < \Delta + 1 = \deg_G y$. In either case, no two neighbors of v have the same degree and so G is highly irregular. $\square$

Next, we determine the existence of highly irregular graphs of a given order. To do this, we first present the following result.

Proposition 2.3 *If G is a highly irregular graph with maximum degree $\Delta \geq 2$, then G contains an induced path of order 4 whose interior vertices have degree Δ in G and whose end-vertices have degree 1 in G.*

Proof By Observation 2.1, the graph G contains two adjacent vertices u and v of degree Δ such that each of u and v is adjacent to a vertex of degree 1 in G. Suppose that u is adjacent to the vertex u_1 of degree 1 and v is adjacent to the vertex v_1 of degree 1. Since $\deg_G u_1 = 1$, the vertex u_1 is adjacent to neither v nor v_1. Similarly, v_1 is not adjacent to u. Thus, (u_1, u, v, v_1) is an induced path of order 4 with the desired property. $\square$

The following is then a consequence of Observation 2.1 and Propositions 2.1 and 2.3.

Proposition 2.4 *For $n = 3, 5, 7$, there is no highly irregular graph of order n.*

Theorem 2.2 *For every integer $n \geq 2$ different from 3, 5, or 7, there is a highly irregular graph of order n.*

Proof Since P_2, P_4, and the graph of order 6 in Fig. 2.3 are highly irregular, it remains to show that there is a highly irregular graph G of order n for each integer $n \geq 8$.

First, suppose that $n = 2\Delta \geq 8$. Let G be the bipartite graph of order n with partite sets $U = \{u_1, u_2, \ldots, u_\Delta\}$ and $W = \{w_1, w_2, \cdots, w_\Delta\}$ such that each

Fig. 2.3 A highly irregular
graph of order 6

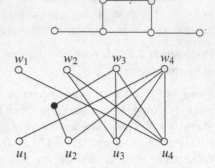

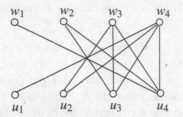

Fig. 2.4 Highly irregular graphs of order 8 and 9

Fig. 2.5 A non-bipartite
highly irregular graph of
order $n = 2\Delta$

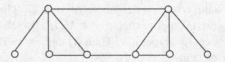

vertex u_i $(1 \leq i \leq \Delta)$ is adjacent to all vertices w_j with $i + j \geq \Delta + 1$. This
is illustrated in Fig. 2.4 for $n = 8$. Hence, $\deg u_i = \deg w_i = i$ for $1 \leq i \leq \Delta$ and
so G is a highly irregular graph of order n. Next, by subdividing the edge $u_2 w_{\Delta-1}$
of G exactly once, we obtain a highly irregular graph G' of order $n + 1 = 2\Delta + 1$.
This is illustrated in Fig. 2.4 for $n = 9$. Consequently, there is a highly irregular
graph of every order $n \geq 8$.   □

We saw in Chap. 1 that if G_n is the unique connected antiregular graph of order
$n \geq 2$, then G_n contains exactly one vertex of degree i for every integer i with $1 \leq$
$i \leq n - 1 = \Delta(G_n)$ with one exception, namely if $i = \lfloor n/2 \rfloor$. Also, we saw in the
proof of Theorem 2.2 that for every even positive integer $n = 2\Delta$, there is a highly
irregular bipartite graph G of order n containing exactly two vertices of degree i for
every integer i with $1 \leq i \leq n/2 = \Delta(G)$. In general, there are examples of such
graphs that are not bipartite. The non-bipartite graph G of order $n = 8$ in Fig. 2.5 is
highly irregular and $\Delta(G) = 4$.

The proof of Theorem 2.2 also shows that for every odd positive integer $n =$
$2\Delta + 1 \geq 9$, there is a highly irregular graph G' of order n containing exactly two
vertices of degree i for every integer i with $1 \leq i \leq (n - 1)/2 = \Delta(G')$ with one
exception, namely there are three vertices of degree 2. There are highly irregular
graphs H of order $n = 2\Delta + 1$ that show the exceptional degree can in fact be
any even positive integer. To see this, let $2k$ be any even positive integer. We begin
with the highly irregular bipartite graph G of order $12k - 4$ and $\Delta(G) = 6k - 2$
described in the proof of Theorem 2.2 with partite sets $U = \{u_1, u_2, \ldots, u_\Delta\}$ and
$W = \{w_1, w_2, \cdots, w_\Delta\}$ where $u_i w_j \in E(G)$ if $i + j \geq \Delta + 1$ so that $\deg u_i =$
$\deg w_i = i$ for $1 \leq i \leq \Delta$. Let H be the graph of order $12k - 3$ obtained from G by
deleting the k edges $u_i w_{6k-i-1}$ for $2k \leq i \leq 3k - 1$ from G, adding a new vertex v,
and adding the edges vu_i, vw_{6k-i-1} for $2k \leq i \leq 3k - 1$. The graph H is highly

irregular with $\Delta(H) = 6k - 2$ containing exactly two vertices of degree i for every integer i with $1 \leq i \leq 6k - 2$ except having three vertices of degree $2k$.

The highly irregular graphs of order $n \geq 8$ constructed in the proof of Theorem 2.2 have the property that $n = 2\Delta(G)$ or $n = 2\Delta(G) + 1$. In fact, it was shown in [2] that every pair n, Δ of integers with $n \geq 2\Delta \geq 8$ is realizable as the order and maximum degree of some highly irregular graph.

Theorem 2.3 *For each pair n, Δ of integers with $n \geq 2\Delta \geq 8$, there exists a highly irregular graph of order n and maximum degree Δ.*

The results above might suggest that there is a high percentage of graphs that are highly irregular, but such is not the case, as was shown in [2].

Theorem 2.4 *For a positive integer n, let $HI(n)$ denote the number of non-isomorphic highly irregular graphs of order n and let $G(n)$ denote the total number of non-isomorphic graphs of order n. Then*

$$\lim_{n \to \infty} \frac{HI(n)}{G(n)} = 0.$$

Obviously, complete graphs of order 3 or more are far from being highly irregular. Indeed, graphs with a relatively large size are very likely not to be highly irregular. The following result gives an upper bound for the size of a highly irregular graph in terms of its order.

Theorem 2.5 *If G is a highly irregular graph of order n and size m, then*

$$m \leq \frac{n(n+2)}{8}. \tag{2.1}$$

Furthermore, the upper bound for m in (2.1) is sharp when n is even.

Proof By Proposition 2.1, if G is a highly irregular graph of order n and maximum degree Δ, then $\Delta \leq n/2$ and G contains at least two vertices of each of the degrees $1, 2, \ldots, \Delta$. First, suppose that G has even order $n = 2r$ and size m. Since $\Delta \leq r$, it follows that

$$m \leq \sum_{i=1}^{r} i = \binom{r+1}{2} = \frac{n(n+2)}{8}.$$

If G has odd order $n = 2r + 1$ and size m, then

$$m \leq \frac{r}{2} + \binom{r+1}{2} = \frac{r(r+2)}{2} = \frac{n(n+2) - 3}{8}.$$

Since the bipartite graph G of order $n = 2\Delta$, where $\Delta(G) = \Delta$, described in the proof of Theorem 2.2 has exactly two vertices of each of the degrees $1, 2, \ldots, \Delta$, it follows that the size of G is

$$m = \sum_{i=1}^{\Delta} i = \binom{\Delta + 1}{2} = \frac{n(n + 2)}{8}.$$

Thus, the upper bound for the size m in (2.1) is sharp when n is even.   □

While Theorem 2.5 provides an upper bound for the size of a highly irregular graph in terms of its order, the following result in [11] gives the minimum size of a highly irregular graph of a given order.

Theorem 2.6 *Let n be a positive integer with $n \neq 3, 5, 7$. The minimum size m_n of a connected highly irregular graph of order n is*

$$m_n = \begin{cases} n - 1 & \text{if } n \neq 6, 11, 12, 13 \\ n & \text{if } n = 6, 12, 13 \\ n + 1 & \text{if } n = 11. \end{cases}$$

Many highly irregular graphs have been shown to have other properties of interest. For example, it was shown in [2] that (i) every graph of order $n \geq 2$ is an induced subgraph of a highly irregular graph of order $4(n - 1)$ and (ii) the smallest order of a highly irregular graph with clique number $\omega \geq 3$ is $4(\omega - 1)$. Highly irregular bipartite graphs and k-chromatic graphs were studied further in [1]. An upper bound for the size of a highly irregular graph of odd order was established in [11, 13]. Additional information regarding highly irregular graphs can also be found in [5, 6, 10, 12].

2.2   Link-Regular Graphs

In a highly irregular graph G, the vertices in the neighborhood of every vertex of G have distinct degrees. The adjacency or non-adjacency of the neighbors of a vertex play no role in whether G is highly irregular. It is possible, however, that the subgraphs induced by the neighborhoods of the vertices of G are distinct regardless of what the degrees of these neighbors may be. This will be the topic of the next section. For the current section, however, we consider graphs where the subgraph induced by the neighborhood of every vertex is the same graph. Necessarily, graphs with this property must be regular. Furthermore, if the graph in question is vertex-transitive (for every two vertices u and v, there is an automorphism mapping u to v), then this graph will necessarily have this property.

The *link* $L(v)$ of a vertex v in a graph G is the subgraph induced by the set of neighbors of v in G, that is, $L(v) = G[N(v)]$. A graph G is called a *link-regular graph* (or a *graph of constant link*) if there exists a graph H such that $L(v) \cong H$ for every vertex v of G. In this case, the graph G is said to be *H-link-regular* and the graph H is called a *link graph*. Thus, a graph H is a *link graph* if there exists an H-link-regular graph G. Furthermore, if G is an H-link-regular graph and H has order r, then G must be r-regular.

This concept was suggested by the Russian mathematician Zykov [15], author of the first textbook in graph theory written in Russian. At the Symposium in Smolenice on the Theory of Graphs and Its Applications, which took place during 17–20 June 1963, Zykov presented the following problem, namely Problem #30, which appeared in the proceedings of this conference.

> **Problem # 30**: Given a finite graph H, does there exist a nonempty G with all neighbourhoods of its vertices isomorphic to H?

While every vertex-transitive graph is a link-regular graph, there are link-regular graphs that are not vertex-transitive. For example, the two 3-regular graphs G_1 and G_2 shown in Fig. 2.6 are not vertex-transitive but are link-regular, where $L(v) = \overline{K_3}$ for each vertex v of G_1 and $L(v) = K_2 + K_1$ for each vertex v of G_2. Consequently, both $\overline{K_3}$ and $K_2 + K_1$ are link graphs.

Several familiar classes of graphs are known to be link graphs. For example, every complete graph is a link graph since K_n is the link of every vertex of K_{n+1} for each positive integer n. Also, every empty graph is a link graph since $\overline{K_n}$ is the link of every vertex of the regular complete bipartite graph $K_{n,n}$. Indeed, for each integer $r \geq 2$, every r-regular triangle-free graph is $\overline{K_r}$-link-regular. More generally, every regular complete multipartite graph is a link graph. For example, $K_{r,r,r}$ is the link of every vertex of the graph $K_{r,r,r,r}$ for each positive integer r.

Since there is a K_3-link-regular graph, namely K_4, there is a C_3-link-regular graph. Also, there is a C_4-link-regular graph since $C_4 = K_{2,2}$ and $K_{2,2,2}$ is a $K_{2,2}$-link-regular graph. In fact, it was shown by Brown and Connelly in [4] that there is a C_n-link-regular graph for each integer $n \geq 3$.

Theorem 2.7 *For each integer $n \geq 3$, there is a C_n-link-regular graph.*

Since $P_2 = K_2$, there is a P_2-link-regular graph. However, there is no P_3-link-regular graph. Clearly, $P_3 = K_{1,2}$. In fact, there is no $K_{1,t}$-link-regular graph for each integer $t \geq 2$ as we show next.

Theorem 2.8 *For each integer $t \geq 2$, no graph is $K_{1,t}$-link-regular.*

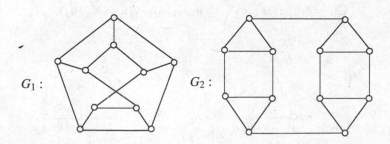

Fig. 2.6 Two link-regular graphs

Proof Assume, to the contrary, that there is a $K_{1,t}$-link-regular graph G for some integer $t \geq 2$. Then G is $(t + 1)$-regular and the link of each vertex of G is $K_{1,t}$. Let $v \in V(G)$. Then $L(v) \cong K_{1,t}$. We may assume that $V(L(v)) = N(v) = \{u, u_1, u_2, \ldots, u_t\}$ and u is the center of $L(v)$. Now, $L(u) \cong K_{1,t}$ where $N(u) = \{v, u_1, u_2, \ldots, u_t\}$ and v is the center of $L(u)$. We now consider $L(u_1)$. Since $L(u_1) \cong K_{1,t}$ and u_1 is adjacent to u and v, it follows that exactly one of u and v is the center of $L(u_1)$, say u is the center of $L(u_1)$. Since u_1 is adjacent to none of $u_2, u_3, \ldots, u_t$, it follows that u_1 must be adjacent to $t-1$ vertices $w_1, w_2, \ldots, w_{t-1}$, where $\{u_2, u_3, \ldots, u_t\} \cap \{w_1, w_2, \ldots, w_{t-1}\} = \emptyset$. However then, u must also be adjacent to all of $w_1, w_2, \ldots, w_{t-1}$ and so $\deg_G u \geq 2t > t + 1$, which is impossible. □

According to Theorem 2.8, the graph $K_{1,4}$ is not a link graph. This graph has order 5 and maximum degree 4. Of course, K_5 also has order 5 and maximum degree 4; however, K_5 is a link graph. This brings up the question as to which graphs of order 5 with maximum degree 4 are link graphs and which are not. By adding one edge or two adjacent edges to $K_{1,4}$, we arrive at the two graphs H_1 and H_2 shown in Fig. 2.7, each of which also fails to be a link graph.

If, on the other hand, one were to add two nonadjacent edges to $K_{1,4}$, we obtain the well-known friendship graph F_2. In general, for each positive integer k, the graph $F_k = kK_2 \vee K_1$ is called a *friendship graph*. Thus, the order of F_k is $n = 2k + 1$ and the maximum degree of F_k is $n - 1 = 2k$ for each positive integer k. The friendship graphs $F_1 = K_3$, F_2, and F_3 are shown in Fig. 2.8. If each vertex in a friendship graph represents a person and each edge represents friendship between the corresponding two people, then every two people have exactly one friend in common. The friendship graph itself has the property that for every two vertices u and v, there is a unique $u - v$ path (u, w, w) of length 2, where the intermediate vertex w is the common friend of u and v. It turns out that every friendship graph is a link graph.

Theorem 2.9 *For each positive integer k, the friendship graph F_k is a link graph.*

Proof We have already seen that $F_1 = K_3$ is a link graph, so we may assume that $k \geq 2$. To construct a $(2k + 1)$-regular graph of order $4k$ that is F_k-link-regular, we begin with the $4k$-cycle

$$C_{4k} = (u_1, u_1', u_2, u_2', \ldots, u_{2k-1}, u_{2k-1}', u_{2k}, u_{2k}', u_1).$$

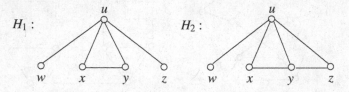

Fig. 2.7 Two graphs of order 5 with maximum degree 4

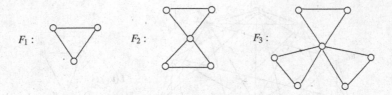

Fig. 2.8 The friendship graphs F_1, F_2 and F_3

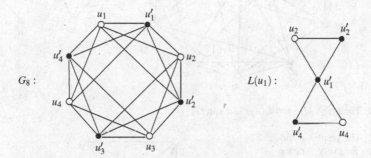

Fig. 2.9 The graph G_8 and $L(u_1) \cong F_2$ in G_8

- If k is even and so $4k \equiv 0 \pmod 8$, then let G_{4k} be the graph obtained from C_{4k} by joining u_i and u'_i to the vertices $u_{i+2\ell-1}$ and $u'_{i+2\ell-1}$ and also to the vertices $u_{i-2\ell+1}$ and $u'_{i-2\ell+1}$ for $\ell = 1, 2, \ldots, k/2$ for each integer i with $1 \le i \le 2k$. Since G_{4k} is vertex-transitive and $L(u_1) \cong F_k$, it follows that G_{4k} is F_k-link-regular. The graph $G_{4k} = G_8$ is shown in Fig. 2.9 for $k = 2$ as well as the subgraph $L(u_1) \cong F_2$ of G_{4k}.
- If k is odd and so $4k \equiv 4 \pmod 8$, then in this case, we let G_{4k} be the graph obtained from C_{4k} by joining u_i and u'_i to the vertices $u_{i+2\ell-1}$ and $u'_{i+2\ell-1}$ and to the vertices $u_{i-2\ell+1}$ and $u'_{i-2\ell+1}$ for $\ell = 1, 2, \ldots, (k-1)/2$ for each integer i with $1 \le i \le 2k$ as well as the vertices u_{i+k} and u'_{i+k}. Since G_{4k} is vertex-transitive and $L(u_1) \cong F_k$, it follows that G_{4k} is F_k-link-regular. The graph $G_{4k} = G_{12}$ is shown in Fig. 2.10 for $k = 3$ as well as the subgraph $L(u_1) \cong F_3$ of G_{4k}.   □

By Theorem 2.8, there is no P_3-link-regular graph. There is, however, a P_4-link-regular graph. For example, the vertex-transitive, 4-regular graph C_8^2 is P_4-link-regular. (The graph C_8^2 is constructed by adding an edge uv to C_8 for every two vertices u and v on C_8 for which $d(u, v) = 2$.) This graph is shown in Fig. 2.11 as well as the link $L(v) \cong P_4$ for every vertex v of C_8^2. Thus, P_4 is a link graph.

Each path belongs to a well-known class of graphs. A forest F is called a *linear forest* if each component of F is a path. All linear forests that are link graphs were determined in [4].

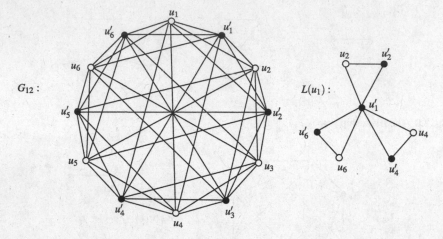

Fig. 2.10 The graph G_{12} and $L(u_1) \cong F_3$ in G_{12}

Fig. 2.11 The graph C_8^2 and
$L(v) \cong P_4$ for every vertex v
in C_8^2

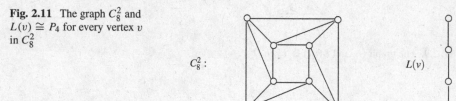

Fig. 2.12 A starlike tree T

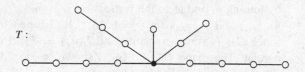

Theorem 2.10 *Let F be a linear forest, where n_i of the paths have length i. Then F is a link graph if and only if*

$$n_2 \leq n_1 + \sum_{i=4}^{\infty} (i-3)n_i.$$

A tree T is *starlike* if T is obtained by subdividing the edges of a star. The branches of T at the center of the star are the *arms* of T. For example, the tree T shown in Fig. 2.12 is starlike. This tree T is obtained from the star $K_{1,5}$ and has five arms, two of length 4 and one of each length 1, 2, and 3. All starlike trees that are link graphs were also determined in [4].

Theorem 2.11 *Let T be a starlike tree where n_i of its arms have length i. Then T is a link graph if and only if*

$$n_2 \le n_1 + \sum_{i=4}^{\infty} (i-3)n_i$$

$$n_1 \le \sum_{i=2}^{\infty} n_i$$

$$2n_1 \le 2n_2 + 3n_3 + \sum_{i=4}^{\infty} (i-3)n_i.$$

Link graphs that are unions of stars (galaxies) or unions of paths and cycles were studied in [9]. Blass et al. [3] investigated trees of small order and determined which of these trees are link graphs and which are not.

In addition to the friendship graphs, another well-known class of graphs is that of the Kneser graphs. For positive integers k and n with $n > 2k$, the *Kneser graph* $KG_{n,k}$ is that graph whose vertices are the k-element subsets of $[n]$ and where two vertices (k-element subsets) A and B are adjacent if and only if A and B are disjoint. Consequently, the Kneser graph $KG_{n,1}$ is the complete graph K_n and the Kneser graph $KG_{5,2}$ is isomorphic to the Petersen graph. Since the Kneser graph $KG_{n+k,k}$ is $KG_{n,k}$-link-regular for every two positive integers k and n with $n > 2k$, it follows that every Kneser graph is a link graph. In particular, the 10-regular Kneser graph $KG_{7,2}$ of order 21 is $KG_{5,2}$-link-regular. Therefore, the Petersen graph P is a link graph and $KG_{7,2}$ is P-link-regular. Hall [7] showed that only two other graphs are P-link-regular.

Theorem 2.12 *For the Petersen graph P, there are exactly three non-isomorphic graphs that are P-link-regular.*

2.3 Link-Irregular Graphs

This topic brings up the question of the existence of link-irregular graphs. A graph G is *link-irregular* if the links of every two vertices of G are not isomorphic. Even though no graph is irregular, there are link-irregular graphs. A link-irregular graph G of order 6 is shown in Fig. 2.13. By adding a new vertex and joining it to a vertex of degree 4 in the graph G of Fig. 2.13, we obtain a link-irregular graph of order 7.

Since a 2-regular graph is a union of cycles, there is no 2-regular link-irregular graph. There are also no 3-regular or 4-regular link-irregular graphs.

Proposition 2.5 *There exists no 3-regular link-irregular graph.*

Proof Assume, to the contrary, that there exists a 3-regular link-irregular graph G. Since there are exactly four non-isomorphic graphs of order 3, it follows that G is a 3-regular graph of order 4. This is impossible since K_4 is the only 3-regular graph of order 4 and K_4 is K_3-link-regular. $\square$

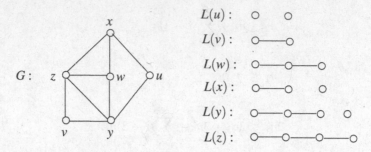

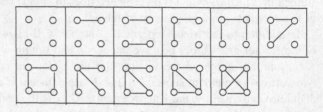

Fig. 2.13 A link-irregular graph of order 6

Fig. 2.14 Eleven non-isomorphic graphs of order 4

Theorem 2.13 *There exists no 4-regular link-irregular graph.*

Proof Assume, to the contrary, that G is a 4-regular link-irregular graph. Since there are exactly eleven non-isomorphic graphs of order 4 (see Fig. 2.14), it follows that G is a 4-regular graph of order $n \leq 11$. There are exactly four graphs of order 4 with maximum degree 3. We claim that if H is a graph of order 4 with maximum degree 3, then H cannot be the link of a vertex of G.

Assume, to the contrary, that there is a graph H of order 4 with maximum degree 3 such that $L(v) \cong H$ for some vertex v of G. Suppose that $N_G(v) = \{u, x, y, z\}$ where u is a vertex of degree 3 in H. This implies that $N_G(u) = \{v, x, y, z\}$. Since both u and v are adjacent to all of x, y, z in G, it follows that $L(u) \cong L(v)$, which is impossible. Hence, as claimed, no graph of order 4 with maximum degree 3 is the link of any vertex of G. Therefore, $n \leq 7$ and so G is a 4-regular graph of order 7 or less. Since both K_5 and $K_{2,2,2}$ are link-regular, it follows that $G \neq K_5$ and $G \neq K_{2,2,2}$. Hence, G is a 4-regular graph of order 7 and so its complement $\overline{G}$ is a 2-regular graph of order 7. Therefore, either $G = \overline{C_7}$ or $G = \overline{C_3 + C_4} = \overline{K_3} \vee 2K_2$. Neither is link-irregular. Consequently, no 4-regular link-irregular graph exists.   □

As a consequence of the results above, it follows that if there is an r-regular link-irregular graph, then $r \geq 5$. Any such graph, however, must have a single automorphism. Regular graphs with only one automorphism are clearly rare. One of the best known of these is the *Frucht graph* shown in Fig. 2.15 (named for Roberto Frucht), which is a 3-regular graph of order 12. Since there is no 3-regular link-irregular graph, the Frucht graph is clearly not such a graph. In fact, nine vertices of this graph have $K_2 + K_1$ as their link while the other three have $\overline{K}_3$ as their link.

Fig. 2.15 The Frucht graph

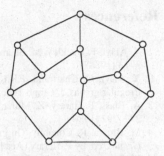

The Frucht graph is also a *Halin graph* (named for Rudolf Halin) since removing the edges of the boundary cycle of the exterior region produces a tree of order 4 or more with no vertices of degree 2. Consequently, it appears that the following conjecture is almost certainly true.

Conjecture 2.1 There exists no regular link-irregular graph.

Evidently, the concept of link-irregular graphs has not been a common topic of study. Among the problems on this topic for which a solution is more likely are the following, the first two of which are suggested by the link-irregular graph of Fig. 2.13.

Problem 2.1 Give examples of link-irregular graphs having minimum degree at least 3.

Problem 2.2 Give examples of link-irregular graphs whose degree set consists of only two elements.

Problem 2.3 Give an example of an infinite class of link-irregular graphs.

Problem 2.4 Given a set S of non-isomorphic graphs of the same order r, does there exist an r-regular graph G such that $\{L(v) : v \in V(G)\} = S$?

For example, if S is the set of the four non-isomorphic graphs of order 3, does there exist a cubic graph G such that $\{L(v) : v \in V(G)\} = S$? If such a graph G exists, then what is the smallest order of G? Furthermore, suppose that $n = 4t$ for some positive integer t. Does there exist a cubic graph G of order n such that each of the four graphs of order 3 is the link of exactly t vertices of G. There are also more general questions dealing with this concept.

Problem 2.5 Let $H_1, H_2, \ldots, H_k$ be $k \geq 2$ non-isomorphic graphs with a prescribed property. Does there exist a graph G such that $\{L(v) : v \in V(G)\} = \{H_1, H_2, \ldots, H_k\}$?

Problem 2.6 For which pairs k, n of positive integers, does there exist a graph G of order n such that the vertices of G have exactly k non-isomorphic links?

References

1. Y. Alavi, F. Buckley, M. Shamula, Highly irregular m-chromatic graphs. Discrete Math. **72**, 3–13 (1988)
2. Y. Alavi, G. Chartrand, F.R.K. Chung, P. Erdős, R.L. Graham, O.R. Oellermann, Highly irregular graphs. J. Graph Theory **11**, 235–249 (1987)
3. A. Blass, F. Harary, Z. Miller, Which trees are link graphs? J. Combin. Theory Ser. B **29**, 277–292 (1980)
4. M. Brown, R. Connelly, On graphs with constant link, in *New Directions in the Theory of Graphs*, ed. by F. Harary (Academic Press, New York, 1973), pp. 19–51
5. M. Brown, R. Connelly, On graphs with a constant link II. Discrete Math. **11**, 199–232 (1975)
6. F. Buckley, Regularizing irregular graphs. Math. Comput. Model. **17**(11), 29–33 (1993)
7. J.I. Hall, Locally Petersen graphs. J. Graph Theory **4**, 173–187 (1980)
8. J.I. Hall, Graphs with constant link and small degree or order. J. Graph Theory **9**, 419–444 (1985)
9. P. Hell, Graphs with given neighbourhoods I, in *Problémes combinatoires et théorie des graphes* (Proc. Colloq. Orsay), Paris (1978), pp. 219–223
10. Z. Majcher, J. Michael, Degree sequences of highly irregular graphs. Discrete Math. **164**, 225–236 (1997)
11. Z. Majcher, J. Michael, Highly irregular graphs with extreme numbers of edges. Discrete Math. **164**, 237–242 (1997)
12. T. Réti, A. Ali, On the comparative study of nonregular networks, in *IEEE 23rd International Conference on Intelligent Engineering Systems*, April 25–27 (Gödöllő, Hungary, 2019), pp. 289–293
13. A. Selvam, Highly irregular bipartite graphs. Indian J. Pure Appl. Math. **27**, 527–536 (1996)
14. B. Zelinka, Graphs with prescribed neighbourhood graphs. Math. Slovaca **35**, 195–197 (1985)
15. A.A. Zykov, Problem 30, in *Theory of Graphs and its Applications (Proc. Symp. Smolenice)* (Prague, 1964), pp. 164–165

Chapter 3
F-Irregular Graphs

In this chapter the concept of irregular graphs is looked at in another way, namely by re-interpreting what is meant by the degree of a vertex.

3.1 F-Degrees

The fundamental observation regarding irregularity in Chap. 1 is:

No nontrivial graph is irregular.

That is, there exists no nontrivial graph the degrees of whose vertices are distinct. This observation depends, of course, on what is meant by the degree of a vertex. This definition is quite standard, however. The degree of a vertex v in a graph G is the number of vertices in G adjacent to v or, equivalently, the number of edges of G incident with v. Also, equivalently, the degree of a vertex v is the number of subgraphs of G isomorphic to K_2 that contain v. Defining the degree of a vertex in this third manner, however, suggests a generalization of this concept (see [2]).

Let F and G be two graphs. The F-*degree* of a vertex v of G, denoted by $F \deg v$, is the number of subgraphs of G isomorphic to F that contain v. Therefore, if $F = K_2$, then $F \deg v = K_2 \deg v$ is the standard degree $\deg v$ of v. The next simplest choice for F is the path P_3 of order 3. To determine the P_3-degree of a vertex v in a graph G, one needs to determine the number of copies of P_3 in which v is an end-vertex and the number of copies of P_3 in which v is the central vertex. The P_3-degree of v is then the sum of these two numbers. The P_3-degrees of the vertices in the graph G of Fig. 3.1 are shown in this figure.

For a graph F, a graph G is F-*regular* if $F \deg u = F \deg v$ for each pair u, v of vertices of G. Of course, F-regular and regular are the same concepts when $F = K_2$. If $F \deg v = s$ for every vertex v of G, then G is F-*regular of degree s*. Furthermore, a graph G is F-regular if and only if all components of G are F-regular of the same degree of regularity. A graph G is F-regular of degree 0 if G contains no subgraph

Fig. 3.1 The P_3-degrees of
the vertices of a graph

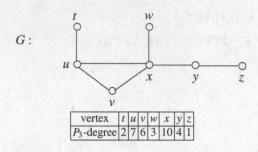

vertex	t	u	v	w	x	y	z
P_3-degree	2	7	6	3	10	4	1

isomorphic to F. Every graph G is G-regular of degree 1. Therefore, P_3 is P_3-regular of degree 1. However, no graph is P_3-regular of degree 2.

Proposition 3.1 *No graph is P_3-regular of degree* 2.

Proof It is sufficient to verify this statement for connected graphs only. Let G be a connected graph of order $n \geq 3$. If $n = 3$ and $G \neq P_3$, then $G = K_3$ and every vertex of K_3 has P_3-degree 3. Thus, we may assume that $n \geq 4$. Suppose first that G has a vertex x of (standard) degree 1 and let y be the neighbor of x and z a neighbor of y. If $\deg y \geq 3$, then $P_3 \deg y \geq 3$; while if $\deg y = 2$, then $P_3 \deg x = 1$. Next, suppose that G contains no vertices of degree 1. Then G contains cycles. However, every vertex on a cycle has P_3-degree 3 or more. Consequently, G is not P_3-regular of degree 2.   □

To determine those integers $s \geq 3$ for which a graph G can be P_3-regular of degree s, we first observe that for each vertex v of G,

$$P_3 \deg v = \binom{\deg v}{2} + \sum_{u \in N(v)} (\deg u - 1). \tag{3.1}$$

Theorem 3.1 *A graph G is P_3-regular of degree s for some integer $s \geq 3$ if and only if G is r-regular for some integer $r \geq 2$. Furthermore, $s = 3\binom{r}{2}$.*

Proof First, assume that G is r-regular where $r \geq 2$. For each vertex v of G, it follows by (3.1) that

$$P_3 \deg v = \binom{r}{2} + \sum_{u \in N(v)} (r - 1) = 3\binom{r}{2}.$$

Thus, G is P_3-regular of degree $s = 3\binom{r}{2}$. For the converse, suppose that G is P_3-regular of degree s where $s \geq 3$. Let u and v be vertices of G such that $\deg u = \delta(G) = \delta$ and $\deg v = \Delta(G) = \Delta$. By (3.1),

$$s = P_3 \deg u \leq \binom{\delta}{2} + \delta(\Delta - 1)$$

and

$$s = P_3 \deg v \geq \binom{\Delta}{2} + \Delta(\delta - 1).$$

Consequently,

$$\binom{\Delta}{2} + \Delta(\delta - 1) \leq \binom{\delta}{2} + \delta(\Delta - 1)$$

and so $(\Delta - \delta)(\Delta + \delta) \leq 3(\Delta - \delta)$. If $\Delta = \delta = r$, then G is r-regular and $s = 3\binom{r}{2}$. If $\Delta \neq \delta$, then $\Delta + \delta \leq 3$ and so $\Delta = 2$ and $\delta = 1$. However then, $G = P_n$, which is not P_3-regular. $\square$

3.2   *F*-Irregularity

For graphs F and G, the graph G is *F-irregular* if the vertices of G have distinct F-degrees. Since K_2-irregular is synonymous with irregular, the following is immediate.

Observation 3.2 *No nontrivial graph is K_2-irregular.*

The next question therefore is: Does there exist a P_3-irregular graph? Actually, we have already answered this question as the graph G of Fig. 3.1 is P_3-irregular. In fact, the graph $G - w$ is also P_3-irregular. Not only are G and $G - w$ both P_3-irregular unicyclic graphs (connected graphs having a unique cycle), it was stated in [2] that these two graphs are the only P_3-irregular unicyclic graphs. There are, however, no nontrivial P_3-irregular trees.

Theorem 3.3 *No nontrivial tree is P_3-irregular.*

Proof Assume, to the contrary, that there exists a nontrivial P_3-irregular tree T. Since no path or star is P_3-irregular, it follows that T is not a path and has diameter $k \geq 3$. Let $P = (v_0, v_1, \ldots, v_k)$ be a longest path in T. Then $\deg v_0 = \deg v_k = 1$. Since $P_3 \deg v_0 \neq P_3 \deg v_k$, it follows that $\deg v_1 \neq \deg v_{k-1}$, say $\deg v_1 > \deg v_{k-1}$. Since $\deg v_1 \geq 3$, this implies that v_1 is adjacent to an end-vertex v_0' distinct from v_0. However then, $P_3 \deg v_0 = P_3 \deg v_0'$, which is a contradiction.
                                                                    $\square$

The next simplest choice for a graph F is K_3. An example of a K_3-irregular graph G of order 13 is shown in Fig. 3.2. Here

$$V(G) = \{u_1, u_2, \ldots, u_6\} \cup \{v_1, v_2, \ldots, v_6\} \cup \{w\},$$

where $G[\{u_1, u_2, \ldots, u_6\}] = K_6 - u_5 u_6$, the set $\{v_1, v_2, \ldots, v_6\}$ is an independent set of vertices and w is joined to each vertex in $\{u_1, u_2, \ldots, u_6\} \cup \{v_6\}$. Furthermore, for $1 \leq i \leq 6$, $v_i u_j \in E(G)$ if $j \geq i$. It might be observed that the construction of this graph is very similar to the antiregular graph shown in Fig. 1.3.

A K_3-irregular graph G_0 of order 15 can be obtained from the graph G of Fig. 3.2 by adding two new vertices u_0 and v_0 to G, joining w to u_0, and joining v_0 to u_j for each $j \geq 0$. By referring to this graph G_0 as G and proceeding as above, a K_3-irregular graph can be constructed for every odd order $n \geq 15$.

We now turn to the topic of F-irregular graphs for connected graphs F of order 4. The graph in Fig. 3.3 of order 12 (similar to the graph of Fig. 3.2) is K_4-irregular. The K_4-degree of each vertex of this graph is also shown in that figure. A simpler choice for a connected graph F of order 4 is the star $K_{1,3}$. Figure 3.3 also shows a $K_{1,3}$-irregular graph of order 8 along with the $K_{1,3}$-degrees of its vertices.

The following result was obtained in [2].

Theorem 3.4 *For every integer $n \geq 3$, there is a K_n-irregular graph and a $K_{1,n}$-irregular graph.*

A P_4-irregular graph of order 9 and a $(K_2 + K_1) \vee K_1)$-irregular graph of order 6 are also shown in Fig. 3.3, as well as the F-degrees of their vertices for each $F \in \{P_4, (K_2 + K_1) \vee K_1)\}$. The graphs shown in Fig. 3.3 bring up the following question.

Problem 3.1 For a given connected graph F, what is the smallest order of an F-irregular graph?

A more fundamental question concerns the following conjecture, which was stated in [2].

Conjecture 3.1 For every connected graph F of order 3 or more, there exists an F-irregular graph.

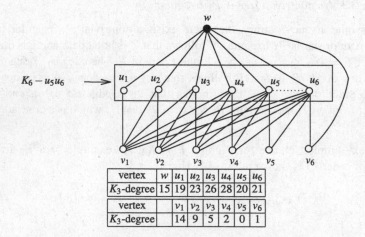

Fig. 3.2 The K_3-degrees of the vertices of a graph

vertex	w	u_1	u_2	u_3	u_4	u_5	u_6
K_3-degree	15	19	23	26	28	20	21

vertex		v_1	v_2	v_3	v_4	v_5	v_6
K_3-degree		14	9	5	2	0	1

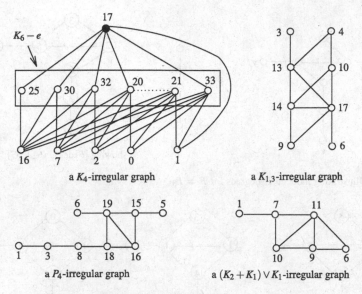

Fig. 3.3 F-irregular graphs for four graphs F of order 4

3.3 Variations of F-Degrees

The topics of F-regularity and F-irregularity discussed in the preceding two sections depend on the concept of the F-degree of a vertex for a given graph F. There are several variations of this concept which lead to related questions dealing with regularity and irregularity. It appears that these topics have not been previously studied.

The K_2-degree $K_2 \deg v = \deg v$ of a vertex v in a graph G may also be defined as the number of induced subgraphs of G isomorphic to K_2 that contain v. This suggests studying the *induced F-degree* $[F] \deg v$ of a vertex v in a graph G, which is different from $F \deg v$ in general if F is not complete. For the graphs G and $F = P_3$ of Fig. 3.4, the induced F-degree $[F] \deg v$ of each vertex v of G is also shown in that figure (placing $[F] \deg v$ inside the vertex v). For the two end-vertices x and y in G, it turns out that $[F] \deg x = 1$ and $[F] \deg y = 3$.

Another variation of the F-degree of a vertex of a graph G is obtained by beginning with a rooted graph F_r, that is, some vertex r of F is selected as the root. Then the *rooted F-degree* $F_r \deg v$ of a vertex v in a graph G is the number of subgraphs of G isomorphic to F where $\phi(r) = v$ in such an isomorphism ϕ. In a similar manner, the *induced rooted F-degree* $[F_r] \deg v$ of a vertex v in a graph G can be defined. For example, consider the graphs G and F of Fig. 3.4. Let r_1 be an end-vertex of F and let r_2 be the vertex of degree 2 in F. For each vertex v of G, the rooted F-degree $F_{r_1} \deg v$, the induced rooted F-degree $[F_{r_1}] \deg v$, and the induced rooted F-degree $[F_{r_2}] \deg v$ of v are shown in Fig. 3.5. Observe that

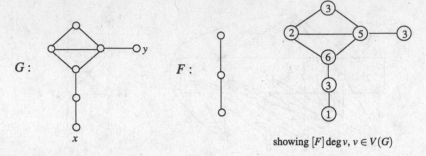

Fig. 3.4 Illustrating induced *F*-degrees for $F = P_3$

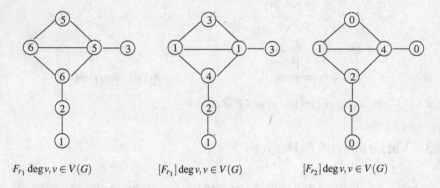

Fig. 3.5 Illustrating rooted *F*-degrees and induced rooted *F*-degrees for $F = P_3$

$[F]\deg v = [F_{r_1}]\deg v + [F_{r_2}]\deg v$ for each vertex v of G. These concepts are related to the concepts of frames and framing number of graphs introduced in [1].

We close this chapter by stating some topics for study.

Problem 3.2 For a non-complete connected graph F, study regularity and irregularity for induced *F*-degrees, rooted *F*-degrees, and induced rooted *F*-degrees.

This problem could also be studied for disconnected graphs F.

References

1. G. Chartrand, H.J. Gavlas, M. Schultz, Framed! A graph embedding problem. Bull. Inst. Combin Appl. **4**, 35–50 (1992)
2. G. Chartrand, K.S. Holbert, O.R. Oellermann, H.C. Swart, *F*-degrees in graphs. Ars Combin. **24**, 133–148 (1987)

Chapter 4
Irregularity Strength

In this chapter, the concept of irregular graphs is looked at in another way, by considering multigraphs rather than graphs or, equivalently, by considering weighted graphs.

4.1 Multigraphs and Weighted Graphs

Once again, we return to the fundamental observation concerning irregularity stated in Chap. 1:

No nontrivial graph is irregular.

Chapter 3 dealt with analyzing this observation when the degree of a vertex in a graph is looked at in a more general manner. Here, we analyze this observation by looking at a graph itself in another manner. The graphs we consider in this book are structures in which every two distinct vertices are joined by at most one edge. These graphs are sometimes called *simple graphs*. There are also structures where two distinct vertices can be joined by more than one edge, although a finite number of edges. In such a structure, often called a *multigraph*, the number of edges incident with a vertex is still referred to as the *degree* of a vertex. The maximum number of edges that join any two distinct vertices in a multigraph has been called the *strength* of the multigraph. If a multigraph has strength 1, then the multigraph is a graph. Figure 4.1 shows a multigraph M of order 3 having strength 3, where $\deg u = 3$, $\deg v = 4$, and $\deg w = 5$.

For each multigraph M, there is an associated graph G called the *underlying graph* of M, which is that graph having the same vertex set as M and where two vertices of G are adjacent if they are joined by at least one edge in M. The underlying graph of the multigraph shown in Fig. 4.1 is therefore K_3.

A. Ali et al., *Irregularity in Graphs*, SpringerBriefs in Mathematics,
https://doi.org/10.1007/978-3-030-67993-4_4

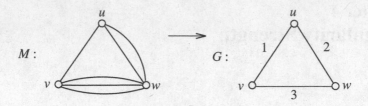

Fig. 4.1 A multigraph with strength 3 and its corresponding weighted graph

As with graphs, a multigraph is *irregular* if distinct vertices have distinct degrees. Unlike graphs, however, irregular multigraphs exist. In fact, the multigraph shown in Fig. 4.1 is irregular. Multigraphs can be looked at in another way. For a multigraph M, let G be the underlying graph of M where an edge uv of G is assigned the label j if u and v are joined by j edges in M. Each such label j is often referred to as the *weight* of the edge uv, resulting in G becoming a *weighted graph*. An assignment of weights to the edges of a graph G is then a weighting of G. A *weighting* of G is therefore a function $w : E(G) \to [k]$ for some positive integer k. The weight of an edge e of G is then denoted by $w(e)$. For the multigraph M of Fig. 4.1, the corresponding weighted graph G is also shown in Fig. 4.1.

In a weighted graph G, the *degree* of a vertex v of G is defined as the sum of the weights of the edges incident with v. This then is the same degree of the vertex v in the corresponding multigraph M. From this we have the following observations, the second of which is a consequence of the first. These two observations can be considered generalizations of Proposition 1.1 and Corollary 1.1.

Observation 4.1 *Let G be a graph with a weighting w, resulting in the weighted graph H. Then*

$$\sum_{v \in V(H)} \deg_H v = 2 \sum_{e \in E(H)} w(e).$$

Observation 4.2 *Every weighted graph has an even number of odd vertices.*

While the multigraph M shown in Fig. 4.1 shows that there is an irregular multigraph of order 3, this brings up the question of the existence of other irregular multigraphs. Certainly, there is no irregular multigraph of order 2. The following result shows that there is no shortage of irregular multigraphs.

Theorem 4.3 *For every connected graph G of order 3 or more, there exists an irregular multigraph whose underlying graph is G.*

Proof Let $E(G) = \{e_1, e_2, \ldots, e_m\}$ and let $w : E(G) \to \mathbb{N}$ be a weighting of G defined by $w(e_i) = 2^{i-1}$ for $1 \le i \le m$. Since no two vertices of G have the same set of edges incident with them, the degrees of every two vertices of the weighted graph G are distinct. Consequently, the corresponding multigraph is irregular. $\qquad\square$

4.2 Irregular Weightings

Theorem 4.3 and its proof suggest a concept that was introduced at the 250th Anniversary of Graph Theory Conference held at Purdue University Fort Wayne in 1986 and in the resulting paper [6]. According to the proof of Theorem 4.3, for every connected graph G of size $m \geq 2$, there is a weighting of G that results in an irregular weighted graph with maximum weight 2^{m-1} or, equivalently, there exists an irregular multigraph of strength 2^{m-1} having G as its underlying graph.

Let G be a connected graph of order 3 or more. A weighting of G that results in an irregular multigraph or irregular weighted graph is an *irregular weighting* of G. The *irregularity strength* $s(G)$ of G is the minimum positive integer k for which there exists an irregular weighting $w : E(G) \rightarrow [k]$ of G. Since no nontrivial graph is irregular, it follows that every connected graph of order at least 3 must have irregularity strength at least 2. There is a familiar class of graphs (seen in Chap. 1) every member of which has irregularity strength 2 (see [5, 17]).

Proposition 4.1 *If G_n is the unique connected antiregular graph of order $n \geq 3$, then $s(G_n) = 2$.*

Proof As we saw, the connected antiregular graph G_n can be described as having vertex set $V(G_n) = \{v_1, v_2, \ldots, v_n\}$ where $v_i v_j \in E(G_n)$ if and only if $i + j \leq n + 1$. Consequently,

$$\deg v_i = \begin{cases} n - i & \text{if } 1 \leq i \leq \lceil n/2 \rceil \\ n + 1 - i & \text{if } \lceil n/2 \rceil + 1 \leq i \leq n. \end{cases} \tag{4.1}$$

Therefore, $\deg v_{\lceil \frac{n}{2} \rceil} = \deg v_{\lceil \frac{n}{2} \rceil + 1} = \lfloor n/2 \rfloor$ and so $v_{\lceil \frac{n}{2} \rceil}$ and $v_{\lceil \frac{n}{2} \rceil + 1}$ are the only two vertices of G_n having the same degree. Let w be the weighting of G_n in which each edge is assigned the weight 2 except for $v_1 v_{\lceil \frac{n}{2} \rceil + 1}$, which is weighted 1. Then the degree $\deg_M v$ of a vertex v in the resulting multigraph M (or weighted graph G_n) is

$$\deg_M v_i = \begin{cases} 2 \deg v_i & \text{if } i \neq 1, \lceil n/2 \rceil + 1 \\ 2 \deg v_i - 1 & \text{if } i = 1, \lceil n/2 \rceil + 1. \end{cases}$$

Thus, the weighted graph G_n is irregular and so $s(G_n) \leq 2$. Hence, $s(G_n) = 2$. □

By the proof of Theorem 4.3, there exists an irregular multigraph of strength 4 having K_3 as its underlying graph. However, we have already seen in Fig. 4.1 that there is an irregular multigraph of strength 3 whose underlying graph is K_3. The irregular weighting of K_3 in Fig. 4.1 therefore shows that $s(K_3) \leq 3$. In fact, not only is $s(K_3) = 3$, but $s(K_n) = 3$ for every integer $n \geq 3$ (see [6]).

Theorem 4.4 *For each integer $n \geq 3$, $s(K_n) = 3$.*

Proof First, let there be given an arbitrary weighting of K_n with the weights 1 and 2, where H is the spanning subgraph of K_n whose edges are weighted 1. Since H is not irregular, H has two vertices u and v of equal degree, which implies that u and v have equal degree in the weighted graph K_n. Consequently, $s(K_n) \geq 3$.

To establish the inequality $s(K_n) \leq 3$, we show that there is an irregular weighting of K_n with the weights $1, 2,$ and 3. Since the weighting of K_3 given in Fig. 4.1 has this property, we may assume that $n \geq 4$. Let G_n be the unique connected antiregular graph of order $n \geq 4$ described in the proof of Proposition 4.1. Thus, $V(G_n) = \{v_1, v_2, \ldots, v_n\}$ whose degrees are given in (4.1). As noted there, these equal degrees are $\lfloor n/2 \rfloor$. Assign the weight 2 to the edges of G_n and the weight 1 to the edges of its complement $\overline{G}_n$. The degrees $\deg_G v_i$ of the vertices v_i in the resulting weighted graph G are then

$$\deg_G v_i = 2 \deg_{G_n} v_i + (n - 1 - \deg_{G_n} v_i) = n - 1 + \deg_{G_n} v_i \qquad (4.2)$$

for $1 \leq i \leq n$. We now increase the weight of each of the edges $v_1 v_2, v_1 v_3, \ldots, v_1 v_{\lceil \frac{n}{2} \rceil}$ by 1, resulting in a weighting w using the weights $1, 2, 3$. By (4.2), this weighting w of K_n results in another weighted graph H. The degrees $\deg_H v_i$ of the vertices v_i in the weighted graph H are then

$$\deg_H v_i = \begin{cases} (2n - 2) + (\lceil n/2 \rceil - 1) & \text{if } i = 1 \\ n + \deg_{G_n} v_i & \text{if } 2 \leq i \leq \lceil n/2 \rceil \\ (n - 1) + \deg_{G_n} v_i & \text{if } \lceil n/2 \rceil + 1 \leq i \leq n. \end{cases}$$

It then follows by (4.1) that

$$\deg_H v_i = \begin{cases} 2n + \lceil n/2 \rceil - 3 & \text{if } i = 1 \\ 2n - i & \text{if } 2 \leq i \leq n. \end{cases}$$

Thus, H is irregular and so $s(K_n) \leq 3$. Consequently, $s(K_n) = 3$.   □

The irregularity strength of many, but not all, complete bipartite graphs has been determined. The following results on $s(K_{p,q})$ where $p \leq q$ were obtained in [12] with the case when $p = q \geq 3$ is odd obtained in [15].

Theorem 4.5 *For integers p and q with $1 \leq p \leq q$,*

$$s(K_{p,q}) = \begin{cases} \left\lceil \frac{q+p-1}{p} \right\rceil & \text{if } q \geq 2p \\ 3 & \text{if } p = q \text{ is even or } 1 < \frac{q}{2} \leq p < q \\ 4 & \text{if } p = q \geq 3 \text{ is odd.} \end{cases}$$

In the case of regular complete multipartite graphs that are not bipartite, the result is stated next (see [12]).

Theorem 4.6 *If G is a regular complete k-partite graph where $k \geq 3$, then $s(G) = 3$.*

The following corollary then summarizes all results on the irregularity strength of regular complete multipartite graphs.

Corollary 4.1 *If G is a regular complete multipartite graph of order at least 3, then*

$$s(G) = \begin{cases} 4 & \text{if } G = K_{r,r} \text{ where } r \geq 3 \text{ is odd} \\ 3 & \text{otherwise.} \end{cases}$$

The irregularity strengths of all paths and cycles have also been determined (see [6, 12]).

Theorem 4.7 *For an integer $n \geq 3$,*

$$s(P_n) = \begin{cases} \frac{n}{2} & \text{if } n \equiv 0 \pmod 4 \\ \frac{n+1}{2} & \text{if } n \text{ is odd} \\ \frac{n+2}{2} & \text{if } n \equiv 2 \pmod 4. \end{cases}$$

Theorem 4.8 *For an integer $n \geq 3$,*

$$s(C_n) = \begin{cases} \frac{n+1}{2} & \text{if } n \equiv 1 \pmod 4 \\ \frac{n+2}{2} & \text{if } n \text{ is even} \\ \frac{n+3}{2} & \text{if } n \equiv 3 \pmod 4. \end{cases}$$

4.3 Bounds

While we have seen that the irregularity strength of graphs belonging to some well-known classes of graphs have been determined, this is certainly not the case for almost all graphs. Bounds, both upper bounds and lower bounds, have been obtained, however, in terms of the order, size, and other numbers associated with a graph.

Lower bounds for the irregularity strength of a connected graph G of order 3 or more can be given in terms of the order and its maximum and minimum degree as well as the number of vertices in G having each degree (see [6]).

Proposition 4.2 *If G is a connected graph of order $n \geq 3$, then*

$$s(G) \geq \frac{n + \delta(G) - 1}{\Delta(G)}.$$

Proof Let there be given an irregular weighting of G with the weights $1, 2, \ldots, s = s(G)$, resulting in the weighted graph G'. Since the degrees of the vertices of G' are distinct elements of the set $S = \{\delta(G), \delta(G) + 1, \ldots, s\Delta(G)\}$ where $|S| = s\Delta(G) - \delta(G) + 1$, it follows that $s\Delta(G) - \delta(G) + 1 \geq n$, giving the desired result.   □

The proof of the following result is similar (see [6]).

Proposition 4.3 *Let G be a connected graph of order $n \geq 3$ with minimum degree $\delta(G)$ and maximum degree $\Delta(G)$ containing n_i vertices of degree i for each integer i with $\delta(G) \leq i \leq \Delta(G)$. Then*

$$s(G) \geq \max \left\{ \frac{n_i - 1}{i} + 1 : \delta(G) \leq i \leq \Delta(G) \right\}.$$

Proof Suppose that $s(G) = s$. Let there be given an irregular weighting of G with the weights $1, 2, \ldots, s$, resulting in the weighted graph G'. If $v \in V(G)$ with $\deg_G v = i$, then the degree $\deg_{G'} v$ of v in G' satisfies $i \leq \deg_{G'} v \leq si$. Hence, each vertex of degree i in G has one of the $si - i + 1 = i(s-1) + 1$ integers in the set $\{i, i + 1, \ldots, si\}$ as its degree in G' and so $n_i \leq i(s-1) + 1$. Therefore, $s(G) = s \geq \frac{n_i - 1}{i} + 1$ for each integer i with $\delta(G) \leq i \leq \Delta(G)$.   □

While the inequality $s(G) \geq n_1$ in Proposition 4.3 is immediate, this lower bound is, nevertheless, useful.

Corollary 4.2 *If G is a connected graph of order 3 or more with n_1 vertices of degree 1, then $s(G) \geq n_1$.*

If the graph G in Proposition 4.3 is regular, then we have the following corollary (see [6]).

Corollary 4.3 *If G is a connected r-regular graph, $r \geq 2$, of order $n \geq 3$, then*

$$s(G) \geq \frac{n - 1}{r} + 1. \tag{4.3}$$

Furthermore, if $n \equiv 2$ (mod 4) or $n \equiv 3$ (mod 4), then

$$s(G) > \frac{n - 1}{r} + 1.$$

Proof Since (4.3) is an immediate consequence of Proposition 4.3, we may assume that $n \equiv 2$ (mod 4) or $n \equiv 3$ (mod 4). We will only consider the case when $n \equiv 2$ (mod 4) since the argument when $n \equiv 3$ (mod 4) is similar. Assume, to the contrary, that $s(G) = s = \frac{n-1}{r} + 1$. Then there is an irregular weighting of G with the weights $1, 2, \ldots, s$, resulting in an irregular weighted graph G'. Hence, each vertex of G' has one of the $sr - r + 1$ degrees $r, r + 1, \ldots, sr$. By assumption, $n = sr - r + 1$ and so the degrees are precisely the n integers $r, r + 1, \ldots, sr$.

Fig. 4.2 An irregular
weighting of the Petersen
graph

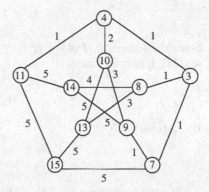

Since $n/2$ of these degrees are odd, G' has an odd number of vertices of odd degree, which contradicts Observation 4.2.   □

By Corollary 4.3, the irregularity strength of the Petersen graph P satisfies $s(P) > \frac{10-1}{3} + 1 = 4$, that is, $s(P) \geq 5$. Since the weighting of the Petersen graph with the weights $1, 2, \ldots, 5$ shown in Fig. 4.2 is irregular, $s(P) \leq 5$ and so $s(P) = 5$.

A lower bound for the irregularity strength of a connected graph of order 3 or more in terms of its order and size was obtained in [6].

Theorem 4.9 *If G is a connected graph of order n and size m with $\Delta(G) \geq 3$, then*

$$s(G) \geq \frac{3n - 2m}{3}.$$

Proof Let $s(G) = s$ and let $w : E(G) \to [s]$ be an irregular weighting of G, resulting in an irregular weighted graph H. Suppose that G has n_i vertices of degree i for $i = 1, 2, \ldots, \Delta = \Delta(G)$. Then

$$n = \sum_{i=1}^{\Delta} n_i \text{ and } 2m = \sum_{i=1}^{\Delta} i n_i.$$

Therefore,

$$3n = 3 \sum_{i=1}^{\Delta} n_i = 2n_1 + n_2 + (n_1 + 2n_2 + 3n_3 + \cdots 3n_\Delta)$$

$$\leq 2n_1 + n_2 + \sum_{i=1}^{\Delta} i n_i = 2n_1 + n_2 + 2m.$$

Thus,

$$3n \leq 2n_1 + n_2 + 2m. \tag{4.4}$$

Since the degrees in H of the $n_1 + n_2$ vertices of degrees 1 and 2 in G belong to the set $[2s]$, it follows that

$$n_1 + n_2 \leq 2s. \tag{4.5}$$

By Corollary 4.2,

$$n_1 \leq s. \tag{4.6}$$

Adding the inequalities in (4.4)–(4.6), we obtain

$$3n + 2n_1 + n_2 \leq 2n_1 + n_2 + 2m + 3s.$$

Consequently, $s(G) \geq \frac{3n-2m}{3}$. □

In the case where the graph in Theorem 4.9 is a tree of order $n \geq 4$ (and size $n - 1$), we have the following corollary.

Corollary 4.4 *If T is a tree of order n with $\Delta(T) \geq 3$, then*

$$s(T) \geq \frac{n+2}{3}.$$

The lower bound for the irregularity strength of a tree given in Corollary 4.4 is sharp as there is an infinite class of trees T of order $n \geq 3$ for which $s(T) = \frac{n+2}{3}$. This is illustrated in Fig. 4.3 for a tree T of order $n = 19$. Since there is an irregular weighting $w : E(T) \rightarrow [7]$, it follows that $s(T) \leq 7$ and so $s(T) = 7$.

It was observed in Corollary 4.2 that if G is a connected graph of order 3 or more containing n_1 vertices of degree 1, then $s(G) \geq n_1$. Therefore, if T is a tree of order 3 or more containing n_1 vertices of degree 1, then $s(T) \geq n_1$. It was shown by Amar and Togni in [2] that equality holds for every tree with $n_2 = 0$.

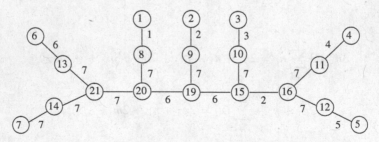

Fig. 4.3 An irregular weighting of a tree T of order 19 with $s(T) = \frac{19+2}{3} = 7$

Theorem 4.10 *If T is a tree of order 3 or more with n_1 vertices of degree 1 and no vertices of degree 2, then $s(T) = n_1$.*

There is, however, a related open problem.

Problem 4.1 Determine the largest integer $n(p)$ such that $s(T) = p$ for every tree T of order $n \leq n(p)$ having p vertices of degree 1.

In the case of unicyclic graphs (connected graphs containing exactly one cycle), Theorem 4.9 has the following corollary.

Corollary 4.5 *If G is a unicyclic graph of order n (and size n), then*

$$s(G) \geq \frac{n}{3}.$$

The lower bound for the irregularity strength of unicyclic graphs given in Corollary 4.5 is also sharp as, here too, there is an infinite class of unicyclic graph G of order n for which $s(G) = \frac{n}{3}$.

There is a general upper bound for the irregularity strength of all connected graphs of order 4 or more (see [1]).

Theorem 4.11 *If G is a connected graph of order $n \geq 4$, then $s(G) \leq n - 1$.*

Since a connected graph G of order $n \geq 3$ and size m has irregularity strength m if and only if G is a star and $m = n - 1$ in this case, the upper bound in Theorem 4.11 is sharp. In the case of trees of order $n \geq 4$ that are not stars, it was also shown in [1, 6] that there is a even better bound.

Theorem 4.12 *If T is a tree of order $n \geq 4$ that is not a star, then $s(T) \leq n - 2$.*

Over the years, many research papers have dealt with the irregularity strength of special classes of graphs. For example, the papers [12–14] deal with the irregularity strength of regular graphs and [2, 4] concern trees. The papers [7, 11] discuss the irregularity strength of dense graphs (those graphs of order n and size m for which m/n is large). The irregularity strength of circulants and grids has been studied in [3] and [8], respectively. Graphs with irregularity strength 2 were studied in [10]. Also, see [5, 16, 17] for more information on this topic.

4.4 Regular Weightings

If a graph G of order at least 3 has a factorization into two factors, one of which is regular and the other factor, say F, is connected, then the irregularity strength of G cannot be any larger than the irregularity strength of F.

Proposition 4.4 *If $\{F_1, F_2\}$ is a factorization of a graph G of order 3 or more where F_1 is connected and F_2 is regular, then $s(G) \leq s(F_1)$.*

Proof Suppose that $s(F_1) = s$. Then there is an irregular weighting $w_1 : E(F_1) \to$ $[s]$ of F_1. Thus, the resulting weighted graph G_1 of F_1 is irregular. Let $w : E(G) \to$ $[s]$ be defined by

$$w(e) = \begin{cases} w_1(e) & \text{if } e \in E(F_1) \\ 1 & \text{if } e \in E(F_2). \end{cases}$$

Let H be the resulting weighted graph of G. Suppose that F_2 is an r-regular graph. Since $\deg_{G_1} u \neq \deg_{G_1} v$ for every two distinct vertices u and v of G, it follows that

$$\deg_H u = \deg_{G_1} u + r \neq \deg_{G_1} v + r = \deg_H v.$$

Thus, H is irregular and so $s(G) \leq s = s(F_1)$.   □

In 1952, Dirac [9] obtained the first theoretical result dealing with Hamiltonian graphs when he proved that if G is a graph of order $n \geq 3$ such that $\delta(G) \geq$ $n/2$, then G is Hamiltonian. The following is a consequence of Dirac's theorem, Theorem 4.8, and Proposition 4.4 (see [12]).

Corollary 4.6 *If G is an r-regular graph of order $n \geq 3$ such that $r \geq n/2$, then*

$$s(G) \leq \left\lceil \frac{n}{2} \right\rceil + 1.$$

By the proof of Proposition 4.4, if G is a connected graph of order 3 or more that can be factored into two factors F_1 and F_2, where F_1 is connected and F_2 is regular, then an irregular weighting of G can be obtained from an irregular weighting of F_1 by assigning the same weight to every edge of F_2.

There is another concept closely related to the irregularity strength of a graph. For a connected graph G of order 3 or more, the *regularity strength* $r(G)$ of G is the minimum positive integer k for which there exists a weighting $w : E(G) \to$ $[k]$ of G such that the resulting weighted graph is regular. If G is a regular graph, then $r(G) = 1$. Unlike the situation for irregularity strength, however, not every connected graph has a regular weighting. For example, no graph of order 3 or more with a pendant edge has a regular weighting and, consequently, no tree of order 3 or more has regularity strength. The regularity strength of the graph F of Fig. 4.4 does exist and, in fact, $r(F) = 2$.

Fig. 4.4 A graph F with
regularity strength 2

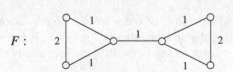

Proposition 4.5 *If $\{F_1, F_2\}$ is a factorization of a graph G of order 3 or more where F_1 is connected and F_2 has a regular weighting for which $r(F_2) \leq s(F_1)$, then $s(G) \leq s(F_1)$.*

Proof Suppose that $s(F_1) = s$ and $r(F_2) = t \leq s(F_1)$. Thus, there is an irregular weighting $w_1 : E(F_1) \rightarrow [s]$ of F_1 and a regular weighting $w_2 : E(F_2) \rightarrow [t]$. Therefore, the resulting weighted graph G_1 of F_1 is irregular and the resulting weighted graph G_2 of F_2 is regular, say G_2 is p-regular. Let $w : E(G) \rightarrow [s]$ be defined by

$$w(e) = \begin{cases} w_1(e) & \text{if } e \in E(F_1) \\ w_2(e) & \text{if } e \in E(F_2). \end{cases}$$

Let H be the resulting weighted graph of G. Since G_1 is irregular, it follows for every two distinct vertices u and v of G that

$$\deg_H u = \deg_{G_1} u + p \neq \deg_{G_1} v + p = \deg_H v.$$

Thus, H is irregular and so $s(G) \leq s = s(F_1)$. □

For example, the graph G of Fig. 4.5 can be factored into two factors F_1 and F_2, where $F_1 \cong P_6$ and F_2 is the graph F of Fig. 4.4. Since $s(F_1) = 4$ by Theorem 4.7 and $r(F_2) = 2 \leq s(F_1)$, it follows by Proposition 4.5 that $2 \leq s(G) \leq 4$. The irregular weighting of G shown in Fig. 4.5 shows that $s(G) \leq 3$.

We show that $s(G) = 3$. Assume, to the contrary, that $s(G) = 2$. Then there exists an irregular weighting $w : (G) \rightarrow [2]$ of G. Since the degree sequence of G is $3, 4, 4, 4, 4, 5$, the vertices of the resulting weighted graph H of G have six distinct degrees in the set $\{3, 4, \ldots, 10\}$. Since H has an even number of odd vertices by Observation 4.2, the degree set of H cannot be $\{3, 4, \ldots, 8\}$. Consequently, some vertex of H must have degree 9 or 10. However, only x can have such a degree, so the degree sequence of H contains either 9 or 10 but not both.

- If $\deg_H x = 10$, then all edges incident with x are weighted 2. Since $w(xy) = 2$, it follows that $\deg_H y \neq 3$, implying that the degree sequence of H must be $4, 5, 6, 7, 8, 10$. In particular, one of the vertices of degree 4 in G must have

Fig. 4.5 An irregular weighting of a graph G

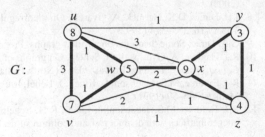

Fig. 4.6 A graph F with
irregularity strength 3

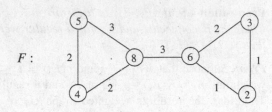

degree 8 in H. If u or z has degree 8 in H, then no vertex of H has degree 4;
while if v or w has degree 8 in H, then no vertex of H has degree 5. This is
impossible.

• If $\deg_H x = 9$, then $\deg_H y = 3$, which implies that no vertex of H has degree 4.
Only v or w can have degree 8 in H but in either case, no vertex of H has
degree 5. This too is impossible.

Therefore, $s(G) \neq 2$ and so $s(G) = 3$.

We saw that the graph F of Fig. 4.4 has regularity strength 2. Since the order
of F is 6, $\delta(F) = 2$, and $\Delta(F) = 3$, it follows by Proposition 4.2 that $s(F) \geq 3$.
The weighting of F in Fig. 4.6 shows that $s(F) \leq 3$ and therefore $s(F) = 3$. This
brings up the following problem.

Problem 4.2 For graphs G where $r(G)$ is defined, determine the relationship
between $s(G)$ and $r(G)$.

References

1. M. Aigner, E. Triesch, Irregular assignments of trees and forests. SIAM J. Discrete Math. **3**,
   439–449 (1990)
2. D. Amar, O. Togni, Irregularity strength of trees. Discrete Math. **190**, 15–38 (1998)
3. J.L. Baril, H. Kheddouci, O. Togni, The irregularity strength of circulant graphs. Discrete Math.
   304, 1–10 (2005)
4. T. Bohman, D. Kravitz, On the irregularity strength of trees. J. Graph Theory **45**, 241–254
   (2004)
5. G. Chartrand, C. Egan, P. Zhang, *How to Label a Graph* (Springer, New York, 2019)
6. G. Chartrand, M.S. Jacobson, J. Lehel, O.R. Oellermann, S. Ruiz, F. Saba, Irregular networks.
   Congr. Numer. **64**, 197–210 (1988)
7. B. Cuckler, F. Lazebnik, Irregularity strength of dense graphs. J. Graph Theory. **58**, 299–313
   (2008)
8. J.H. Dinitz, D.K. Garnick, A. Gyárfás, On the irregularity strength of the $m \times n$ grid. J. Graph
   Theory **16**, 355–374 (1992)
9. G.A. Dirac, Some theorems on abstract graphs. Proc. London Math. Soc. **2**, 69–81 (1952)
10. R.J. Faudree, A. Gyárfás, R.H. Schelp, On graphs of irregularity strength 2, in *Combinatorics,
    Colloq. Math. Soc.* János Bolyai vol. 52 (North Holland, Amsterdam, 1987), pp. 239–246
11. R.J. Faudree, M.S. Jacobson, L. Kinch, J. Lehel, Irregularity strength of dense graphs. Discrete
    Math. **91**, 45–59 (1991)
12. R.J. Faudree, M.S. Jacobson, J. Lehel, R.H. Schelp, Irregular networks, regular graphs and
    integer matrices with distinct row and column sums. Discrete Math. **76**, 223–240 (1989)

13. R.J. Faudree, J. Lehel, Bound on the irregularity strength of regular graphs, in *Combinatorics Colloq. Math. Soc.* János Bolyai, vol. 52 (North Holland, Amsterdam, 1987), pp. 247–256
14. E. Flandrin, A. Marczyk, J. Przybylo, J.F. Saclé, M. Woźniak, Neighbor sum distinguishing index. Graphs Combin. **29**, 1329–1336 (2013)
15. A. Gyárfás, The irregularity strength of $K_{m,m}$ is 4 for odd m. Discrete Math. **71**, 273—274 (1988)
16. J. Lehel, Facts and quests on degree irregular assignments. Semantic Scholar (1988) 765–781
17. P. Zhang, *Color-Induced Graph Colorings* (Springer, New York, 2015)

Chapter 5
Rainbow Mean Index

It was seen in the preceding chapter that every connected graph of order 3 or more has an irregular weighting. Therefore, if G is a connected graph of order 3 or more, there exists a weighting $w : E(G) \rightarrow [k]$ for some integer $k \geq 2$ such that the vertices in the resulting weighted graph H of G have distinct degrees. To obtain an irregular weighted graph H with a given underlying graph G, it may very well be necessary for the degrees of the vertices of H to be large, possibly some much larger than the order of G. With this observation in mind, a different weighting of a graph was introduced, referred to as an edge coloring, which was employed to produce an irregular vertex coloring, with the goal of minimizing the largest vertex color. Since irregular colorings are often called rainbow colorings, this is the terminology we use.

5.1 Rainbow Mean Colorings

Let G be a connected graph of order 3 or more. For a vertex v of G, let E_v denote the set of edges incident with v in G. An edge coloring $c : E(G) \rightarrow \mathbb{N}$ of G, where $\mathbb{N}$ is the set of positive integers, is called a *mean coloring* if the *chromatic mean* of each vertex v, defined by

$$\text{cm}(v) = \frac{\sum_{e \in E_v} c(e)}{\deg v},$$

is an integer. That is, the color of the vertex v is an integer that is the average of the colors of the edges incident with v. If distinct vertices have distinct chromatic means, then the edge coloring c is called a *rainbow mean coloring* of G, a concept introduced and studied in [1] and studied further in [3, 4]. While, initially, it may not be clear which graphs possess such an edge coloring, the following result shows

© The Author(s), under exclusive license to Springer Nature Switzerland AG 2021
A. Ali et al., *Irregularity in Graphs*, SpringerBriefs in Mathematics,
https://doi.org/10.1007/978-3-030-67993-4_5

that for every connected graph of order 3 or more, such an edge coloring always exists.

Theorem 5.1 *Every connected graph of order* 3 *or more has a rainbow mean coloring.*

Proof Suppose that G is a connected graph with $E(G) = \{e_1, e_2, \ldots, e_m\}$ where $m \geq 2$. Thus, $\Delta(G) = \Delta \geq 2$. Let $k = 2\Delta$ and $t = \Delta! k^m$. Define the edge coloring $c : E(G) \to [t]$ by $c(e_i) = \Delta! k^i$ for $1 \leq i \leq m$. We show that the coloring c has the desired property. Assume, to the contrary, that there are two distinct vertices u and v of G such that $\mathrm{cm}(u) = \mathrm{cm}(v)$. Let $\deg u = r$ and $\deg v = s$, where $r \leq s$ say, and let $E_u = \{e_{i_1}, e_{i_2}, \ldots, e_{i_r}\}$ and $E_v = \{e_{j_1}, e_{j_2}, \ldots, e_{j_s}\}$ where $1 \leq i_1 < i_2 < \cdots < i_r \leq m$ and $1 \leq j_1 < j_2 < \cdots < j_s \leq m$. If $uv \notin E(G)$, then $E_u \cap E_v = \emptyset$; while if $uv \in E(G)$, then $E_u \cap E_v = \{uv\}$. Consequently,

$$\mathrm{cm}(u) = \frac{\Delta!}{r} \left(k^{i_1} + k^{i_2} + \cdots + k^{i_r} \right)$$

$$\mathrm{cm}(v) = \frac{\Delta!}{s} \left(k^{j_1} + k^{j_2} + \cdots + k^{j_s} \right),$$

where both $\mathrm{cm}(u)$ and $\mathrm{cm}(v)$ are positive integers. We consider two cases, according to whether $r = s$ or $r < s$.

Case 1 $r = s$. Then $k^{i_1} + k^{i_2} + \cdots + k^{i_r} = k^{j_1} + k^{j_2} + \cdots + k^{j_r}$.

- First, suppose that $i_r \neq j_r$. We may assume that $i_r < j_r$. Let $p = j_r \geq 2$. Since $k = 2\Delta \geq 4$, it follows that $k^p > k + k^2 + \ldots + k^{p-1}$. However then,

$$k^{j_1} + k^{j_2} + \cdots + k^{j_r} \geq k^{j_r} = k^p > k + k^2 + \ldots + k^{p-1} \geq k^{i_1} + k^{i_2} + \cdots + k^{i_r},$$

which is a contradiction.
- Next, suppose that $i_r = j_r$. Then $k^{i_1} + k^{i_2} + \cdots + k^{i_{r-1}} = k^{j_1} + k^{j_2} + \cdots + k^{j_{r-1}}$ and $i_{r-1} \neq j_{r-1}$. We can apply the argument above to produce a contradiction.

Case 2 $r < s$. Then $s \left[k^{i_1} + k^{i_2} + \cdots + k^{i_r} \right] = r \left[k^{j_1} + k^{j_2} + \cdots + k^{j_s} \right]$.

- First, suppose that $i_r < j_s$. Let $p = j_s \geq 2$. Since $1 > \frac{1}{k^{p-1}} + \frac{1}{k^{p-2}} + \cdots + \frac{1}{k}$, it follows that

$$2 > \frac{1}{k^{p-1}} + \frac{1}{k^{p-2}} + \cdots + \frac{1}{k} + 1 > \frac{1}{k^{p-2}} + \frac{1}{k^{p-3}} + \cdots + \frac{1}{k} + 1.$$

Hence, $k = 2\Delta > \Delta \left(\frac{1}{k^{p-2}} + \frac{1}{k^{p-3}} + \cdots + 1 \right)$. Because $\Delta \geq s/r$, it follows that

$$k^{j_1} + k^{j_2} + \cdots + k^{j_s} \geq k^{j_s} = k^p = k(k^{p-1}) > \Delta \left(\frac{1}{k^{p-2}} + \frac{1}{k^{p-3}} + \cdots + 1 \right) k^{p-1}$$

$$= \Delta(k + k^2 + \cdots + k^{p-1}) \geq \frac{s}{r}(k + k^2 + \cdots + k^{p-1})$$

$$\geq \frac{s}{r}\left[k^{i_1} + k^{i_2} + \cdots + k^{i_r}\right],$$

which is a contradiction.

- Next, suppose that $i_r \geq j_s$. The argument in Case 1 shows that

$$k^{i_1} + k^{i_2} + \cdots + k^{i_r} > k^{j_1} + k^{j_2} + \cdots + k^{j_s}.$$

Since $r < s$, it follows that $1 \geq r/s$ and so

$$k^{i_1} + k^{i_2} + \cdots + k^{i_r} > k^{j_1} + k^{j_2} + \cdots + k^{j_s} > \frac{r}{s}\left[k^{j_1} + k^{j_2} + \cdots + k^{j_s}\right],$$

which is a contradiction. □

For a rainbow mean coloring c of a graph G, the maximum vertex color is the *rainbow chromatic mean index* (or simply, the *rainbow mean index*) $\mathrm{rm}(c)$ of c. That is,

$$\mathrm{rm}(c) = \max\{\mathrm{cm}(v) : v \in V(G)\}.$$

The *rainbow chromatic mean index* (or the *rainbow mean index*) $\mathrm{rm}(G)$ of the graph G itself is defined as

$$\mathrm{rm}(G) = \min\{\mathrm{rm}(c) : c \text{ is a rainbow mean coloring of } G\}.$$

Consequently, if G is a connected graph of order $n \geq 3$, then $\mathrm{rm}(G) \geq n$. A primary question here is the following:

For a connected graph G of order $n \geq 3$, how much larger than n can $\mathrm{rm}(G)$ be?

It was an attempt to study this question that led to the introduction of rainbow mean colorings.

For a mean coloring of a connected graph G, the *chromatic sum* $\mathrm{cs}(v)$ of a vertex v of G is the sum of the colors of the edges incident with v. Hence, $\mathrm{cs}(v) = \deg v \cdot \mathrm{cm}(v)$. The following result in [1] is an elementary yet useful result.

Proposition 5.1 *If c is a mean coloring of a connected graph G, then*

$$\sum_{v \in V(G)} \mathrm{cs}(v) = 2 \sum_{e \in E(G)} c(e).$$

Proof When the chromatic sums of the vertices of G are added, the color of each edge xy is counted twice, once in $\mathrm{cs}(x)$ and once in $\mathrm{cs}(y)$. □

In Proposition 5.1, if c is the edge coloring of a graph G such that $c(e) = 1$ for each edge e of G, then $cs(v) = \deg v$ for every vertex v of G and so $\sum_{v \in V(G)} \deg v = 2|E(G)|$, which is the First Theorem of Graph Theory.

Let G be a connected graph of order 3 or more with a mean coloring. A vertex v of G is called *chromatically even* if $cs(v)$ is even and v is *chromatically odd* otherwise. The following result in [1] is an immediate consequence of Proposition 5.1.

Proposition 5.2 *Let G be a connected graph with a mean coloring. Then G has an even number of chromatically odd vertices.*

Proof By Proposition 5.1, the sum of the chromatic sums of all vertices of G is an even number. Therefore, there is an even number of chromatically odd vertices.   □

A consequence of Proposition 5.2 is stated next (see [1]).

Corollary 5.1 *Let G be a connected graph of order $n \geq 6$ where $n \equiv 2 \pmod 4$ such that all vertices of G are odd. Then $\mathrm{rm}(G) \geq n + 1$.*

Proof Assume, to the contrary, that $\mathrm{rm}(G) = n$. Since $n \equiv 2 \pmod 4$ and $n \geq 6$, it follows that $n = 4k + 2$ for some positive integer k. Hence, G has $2k + 1$ chromatically odd vertices. This contradicts Proposition 5.2.   □

For example, the Petersen graph P is a connected cubic graph of order $10 \equiv 2$ (mod 4). Figure 5.1 shows a rainbow mean coloring c of P with $\mathrm{rm}(c) = 11$. Thus, $\mathrm{rm}(P) = 11$ by Corollary 5.1.

For examples of the concepts we have described, we consider the rainbow mean index of paths and cycles. First, we determine $\mathrm{rm}(P_4)$, which we will see is a special case.

Proposition 5.3 $\mathrm{rm}(P_4) = 5$.

Proof The edge coloring of P_4 in Fig. 5.2 shows that $\mathrm{rm}(P_4) \leq 5$. Next, we show that $\mathrm{rm}(P_4) \geq 5$. Assume, to the contrary, that there is a rainbow mean coloring c of P_4 such that $\mathrm{rm}(c) = 4$. Let $P_4 = (v_1, v_2, v_3, v_4)$. Since $\{cm(v_i) : 1 \leq i \leq 4\} = [4]$, some vertex of P_4 is colored 1 and at least one edge of P_4 is colored 1. This implies that all edges of P_4 are colored with odd integers. Since one vertex of P_4

Fig. 5.1 A rainbow mean coloring of the Petersen graph P

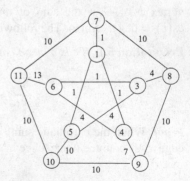

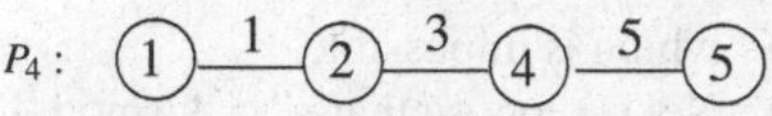

Fig. 5.2 A rainbow mean
coloring of P_4

$$P_4: \quad \textcircled{1} \xrightarrow{\ 1\ } \textcircled{2} \xrightarrow{\ 3\ } \textcircled{4} \xrightarrow{\ 5\ } \textcircled{5}$$

is colored 4, some edge of P_4 (necessarily the interior edge) is colored at least 5. The only possibility is for this edge to be colored 5 and the remaining edge to be colored 3. However then, two vertices are colored 3, which is impossible. □

For all other paths P_n of order $n \geq 3$, their rainbow mean index is n however. In each of these cases, an appropriate rainbow mean coloring can be given.

Theorem 5.2 *For each integer $n \geq 3$ with $n \neq 4$, $\text{rm}(P_n) = n$.*

Since cycles are 2-regular, every rainbow mean coloring of a cycle must assign colors of the same parity to every edge of a cycle. From this, it immediately follows that $\text{rm}(C_3) = 4$. The rainbow mean index of every cycle is stated next.

Theorem 5.3 *For each integer $n \geq 3$,*

$$\text{rm}(C_n) = \begin{cases} n & if\, n \equiv 0, 1 \pmod 4 \\ n+1 & if\, n \equiv 2, 3 \pmod 4 \end{cases}$$

Proof It can be shown that (i) there is a rainbow mean coloring c of C_n with $\text{rm}(c) = n$ where $n \equiv 0, 1 \pmod 4$ and (ii) there is a rainbow mean coloring c of C_n with $\text{rm}(c) = n + 1$ where $n \equiv 2, 3 \pmod 4$. From this, we need only verify that $\text{rm}(C_n) \geq n + 1$ where $n \equiv 2, 3 \pmod 4$. Assume, to the contrary, that there is a rainbow mean coloring c of C_n such that $\text{rm}(c) = n$ for some integer n with $n \equiv 2, 3 \pmod 4$. Let $C_n = (v_1, v_2, \dots, v_n, v_{n+1} = v_1)$ where $e_i = v_i v_{i+1}$ for $1 \leq i \leq n$. Since the color of some vertex of C_n is 1, the colors of the two edges incident with this vertex are both 1. This implies that $c(e)$ is odd for each $e \in E(C_n)$. Thus, $c(e_i) = 2a_i + 1$ for some nonnegative integer a_i for each integer i with $1 \leq i \leq n$.

- First, suppose that $n \equiv 2 \pmod 4$. Then $n = 4k + 2$ for some positive integer k and so

$$2 \sum_{v \in V(C_n)} cm(v) = 2 \binom{4k+3}{2} = (4k+3)(4k+2) = 16k^2 + 20k + 6.$$

Hence, $2 \sum_{v \in V(C_n)} cm(v) \equiv 2 \pmod 4$. On the other hand,

$$2 \sum_{v \in V(C_n)} cm(v) = 2 \sum_{i=1}^{4k+2} c(e_i) = 2 \sum_{i=1}^{4k+2} (2a_i + 1) = \sum_{i=1}^{4k+2} (4a_i + 2)$$

$$= \left[\sum_{i=1}^{4k+2} 4a_i \right] + (8k+4) \equiv 0 \pmod 4,$$

which is impossible.

- Next, suppose that $n \equiv 3 \pmod 4$. We have already observed that $\operatorname{rm}(C_3) = 4$. Thus, $n = 4k + 3$ for some positive integer k. Then

$$2 \sum_{v \in V(C_n)} \operatorname{cm}(v) = 2\binom{4k+4}{2} = (4k+4)(4k+3) = 4(k+1)(4k+3).$$

Hence, $2\sum_{v \in V(C_n)} \operatorname{cm}(v) \equiv 0 \pmod 4$. Furthermore,

$$2 \sum_{v \in V(C_n)} \operatorname{cm}(v) = 2 \sum_{i=1}^{4k+3} c(e_i) = 2 \sum_{i=1}^{4k+3} (2a_i + 1) = \sum_{i=1}^{4k+3} (4a_i + 2)$$

$$= \left[\sum_{i=1}^{4k+3} 4a_i \right] + (8k + 6) \equiv 2 \pmod 4,$$

which is impossible.

Therefore, $\operatorname{rm}(C_n) \geq n + 1$ if $n \equiv 2 \pmod 4$ or $n \equiv 3 \pmod 4$. $\qquad \square$

5.2   Complete Graphs and Complete Bipartite Graphs

We now turn our attention to complete graphs and complete bipartite graphs. In order to determine the rainbow mean index of the graphs in these two classes, we consider a matrix representation of an edge-colored graph. Let G be a connected graph of order $n \geq 3$ with $V(G) = \{v_1, v_2, \ldots, v_n\}$ and let $c : E(G) \to \mathbb{N}$ be an edge coloring of G. The *matrix representation* M of G with the edge coloring c is the $n \times n$ matrix $[m_{i,j}]$ where

$$m_{i,j} = \begin{cases} c(v_i v_j) & \text{if } v_i v_j \in E(G) \\ 0 & \text{otherwise.} \end{cases}$$

There are several elementary observations that can be made about the matrix representation M of a graph G of order n with an edge coloring c. First, all entries along the main diagonal of M are 0 since no vertex of G is adjacent to itself. Second, M is a symmetric matrix. Also, the sum of the entries in row i (equivalently, in column i) is

$$\deg v_i \cdot \operatorname{cm}(v_i) = \operatorname{cs}(v_i) \text{ for } 1 \leq i \leq n.$$

First, we show that the rainbow mean index of a complete graph K_n of order $n \geq 3$ is either n or $n + 1$.

Theorem 5.4 *For an integer $n \geq 3$,*

$$\text{rm}(K_n) = \begin{cases} n & \text{if } n \geq 4 \text{ and } n \equiv 0, 1, 3 \pmod 4 \\ n+1 & \text{if } n = 3 \text{ or } n \equiv 2 \pmod 4. \end{cases}$$

Proof Since $\text{rm}(K_3) = \text{rm}(C_3) = 4$ by Theorem 5.3, we may assume that $n \geq 4$. For $n \geq 4$ and $n \equiv 0, 1, 3 \pmod 4$, it suffices to show that there is a rainbow mean coloring of K_n having rainbow mean index n. We only consider the situation when $n \equiv 0 \pmod 4$ since the argument for the cases when $n \equiv 1, 3 \pmod 4$ is similar. Thus, $n = 4k$ for some positive integer k.

In order to describe a rainbow mean coloring c_n of K_n with $\text{rm}(c_n) = n$, we construct an $n \times n$ symmetric matrix M_n. First, we define, recursively, a sequence $B_1, B_2, \ldots, B_k$ of 4×4 symmetric matrices. For $a = n - 1$, let

$$B = \begin{bmatrix} 0 & a & a & 2a \\ a & 0 & 2a & a \\ a & 2a & 0 & a \\ 2a & a & a & 0 \end{bmatrix} \quad \text{and } B_1 = \begin{bmatrix} 0 & 1 & 1 & 1 \\ 1 & 0 & 1 & a+1 \\ 1 & 1 & 0 & 2a+1 \\ 1 & a+1 & 2a+1 & 0 \end{bmatrix}.$$

For $2 \leq i \leq k$, define $B_i = B_{i-1} + B = B_1 + (i-1)B$. Thus,

$B_i = B_1 + (i-1)B$

$$= \begin{bmatrix} 0 & 1 & 1 & 1 \\ 1 & 0 & 1 & a+1 \\ 1 & 1 & 0 & 2a+1 \\ 1 & a+1 & 2a+1 & 0 \end{bmatrix} + \begin{bmatrix} 0 & (i-1)a & (i-1)a & 2(i-1)a \\ (i-1)a & 0 & 2(i-1)a & (i-1)a \\ (i-1)a & 2(i-1)a & 0 & (i-1)a \\ 2(i-1)a & (i-1)a & (i-1)a & 0 \end{bmatrix}$$

$$= \begin{bmatrix} 0 & (i-1)a+1 & (i-1)a+1 & 2(i-1)a+1 \\ (i-1)a+1 & 0 & 2(i-1)a+1 & ia+1 \\ (i-1)a+1 & 2(i-1)a+1 & 0 & (i+1)a+1 \\ 2(i-1)a+1 & ia+1 & (i+1)a+1 & 0 \end{bmatrix}.$$

To describe the $n \times n$ matrix M_n, we begin with a $k \times k$ matrix $A = [a_{i,j}]$ and then replace the entry $a_{i,i}$ on the main diagonal of A by the 4×4 matrix B_i for $1 \leq i \leq k$ and each entry off the main diagonal of A by the 4×4 matrix J, each of whose entries is 1. That is, $M_n = [M_{i,j}]$ is an $n \times n$ matrix, where $M_{i,j}$ is a 4×4 matrix such that

$$M_{i,j} = \begin{cases} B_i & \text{if } 1 \leq i = j \leq k \\ J & \text{if } 1 \leq i \neq j \leq k. \end{cases}$$

Thus,

$$M_4 = B_1, \quad M_8 = \begin{bmatrix} B_1 & J \\ J & B_2 \end{bmatrix}, \quad \text{and } M_{12} = \begin{bmatrix} B_1 & J & J \\ J & B_2 & J \\ J & J & B_3 \end{bmatrix}.$$

If we were to add the entries in row i (or in column i) in M_n, then we obtain ia for $1 \le i \le n$. That is, if $M_n = [m_{i,j}]$, then

$$\sum_{j=1}^{n} m_{i,j} = ia = i(n-1) \text{ for } 1 \le i \le n. \tag{5.1}$$

We now define an edge coloring $c : E(K_n) \to \mathbb{N}$ by $c(v_i v_j) = m_{i,j}$ for each pair i, j of integers with $1 \le i \le j \le n$ and $i \ne j$. Since $\text{cm}(v_i) = \frac{1}{n-1} \sum_{j=1}^{n} m_{i,j} = i$ for $1 \le i \le n$ by (5.1), it follows that c is a rainbow mean coloring of K_n with $\text{rm}(c) = n$. For example,

$$M_4 = \begin{bmatrix} 0 & 1 & 1 & 1 \\ 1 & 0 & 1 & 4 \\ 1 & 1 & 0 & 7 \\ 1 & 4 & 7 & 0 \end{bmatrix} \quad \text{and } M_8 = \begin{bmatrix} 0 & 1 & 1 & 1 & 1 & 1 & 1 & 1 \\ 1 & 0 & 1 & 8 & 1 & 1 & 1 & 1 \\ 1 & 1 & 0 & 15 & 1 & 1 & 1 & 1 \\ 1 & 8 & 15 & 0 & 1 & 1 & 1 & 1 \\ 1 & 1 & 1 & 1 & 0 & 8 & 8 & 15 \\ 1 & 1 & 1 & 1 & 8 & 0 & 15 & 15 \\ 1 & 1 & 1 & 1 & 8 & 15 & 0 & 22 \\ 1 & 1 & 1 & 1 & 15 & 15 & 22 & 0 \end{bmatrix}$$

The matrices M_4 and M_8 give rise to rainbow mean colorings of K_4 and K_8 as shown in Fig. 5.3, respectively, where each edge drawn with a thin line is colored by 1.

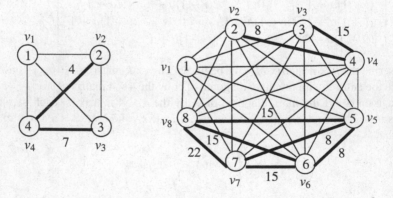

Fig. 5.3 Rainbow mean colorings of K_4 and K_8

Next, suppose that $n \equiv 2 \pmod 4$. By Corollary 5.1, it suffices to show that there is a rainbow mean coloring c_n of K_n with $\mathrm{rm}(c_n) = n+1$. Again, we construct an $n \times n$ symmetric matrix M_n by constructing a sequence $A_1, A_2, \ldots, A_k$ of symmetric matrices, where A_1 is a 6×6 matrix and A_i is a 4×4 matrix for $2 \le i \le k$. For

$$a = n - 1, \text{ recall that } B = \begin{bmatrix} 0 & a & a & 2a \\ a & 0 & 2a & a \\ a & 2a & 0 & a \\ 2a & a & a & 0 \end{bmatrix}. \text{ Define}$$

$$A_1 = \begin{bmatrix} 0 & 1 & 1 & 1 & 1 & 1 \\ 1 & 0 & a+1 & 1 & 1 & 1 \\ 1 & a+1 & 0 & 1 & 1 & a+1 \\ 1 & 1 & 1 & 0 & a+1 & 2a+1 \\ 1 & 1 & 1 & a+1 & 0 & 3a+1 \\ 1 & 1 & a+1 & 2a+1 & 3a+1 & 0 \end{bmatrix} \text{ and } A_2 = \begin{bmatrix} 0 & a & 3a & 3a \\ a & 0 & 3a & 4a \\ 3a & 3a & 0 & 3a \\ 3a & 4a & 3a & 0 \end{bmatrix}.$$

For $3 \le i \le k$, define $A_i = A_{i-1} + B = A_2 + (i-2)B$.

To describe the $n \times n$ matrix M_n, we begin with a $k \times k$ matrix $A = [a_{i,j}]$ and then replace the entry $a_{i,i}$ on the main diagonal of A by the matrix A_i for $1 \le i \le k$ and each entry off the main diagonal of A by the matrix J, each of whose entries is 1. Thus, $a_{1,1}$ is replaced by the 6×6 matrix A_1 and $a_{i,i}$ for $2 \le i \le k$ is replaced by the 4×4 matrix A_i. That is, $M_n = [M_{i,j}]$ is an $n \times n$ matrix where

$$M_{i,j} = \begin{cases} A_i & \text{if } 1 \le i = j \le k \\ J & \text{if } 1 \le i \ne j \le k. \end{cases}$$

Thus, $M_6 = A_1$, $M_{10} = \begin{bmatrix} A_1 & J \\ J & A_2 \end{bmatrix}$ and $M_{14} = \begin{bmatrix} A_1 & J & J \\ J & A_2 & J \\ J & J & A_3 \end{bmatrix}$. We now define a rainbow mean coloring $c : E(K_n) \to \mathbb{N}$ by $c(v_i v_j) = m_{i,j}$ for each pair i, j of integers with $1 \le i \le j \le n$ and $i \ne j$. For example, the matrix M_6 gives rise to the rainbow mean coloring of K_6 as shown in Fig. 5.4, where again each edge drawn with a thin line is colored by 1. Since $\mathrm{rm}(c) = n+1$, it follows that $\mathrm{rm}(K_n) = n+1$ for each integer $n \ge 6$ with $n \equiv 2 \pmod 4$. □

Next, we consider complete bipartite graphs. We begin with a special class of complete bipartite graphs, namely stars. As we saw, the number of vertex colors in a rainbow mean coloring of a graph is at least the order of the graph. Therefore, if the order of G is $n \ge 3$, then the number of vertex colors must be at least n. With all the conditions required for a graph to have such an edge coloring, one might anticipate that for some graphs the largest vertex color may be significantly greater than the order of the graph. However, for each connected graph G of order $n \ge 3$ that we have considered thus far, we have seen that either $\mathrm{rm}(G) = n$ or $\mathrm{rm}(G) = n + 1$. While this observation may suggest a conjecture, the following result dealing with

Fig. 5.4 A rainbow mean
coloring of K_6

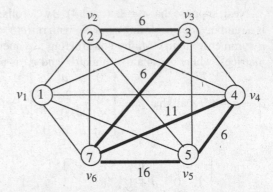

the stars $K_{1,n-1}$ of order $n \geq 3$ indicates that the value of $rm(G)$ for a connected
graph G of order $n \geq 3$ can be one of at least *three* integers rather than only one of
two integers.

Theorem 5.5 *If G is a star of order $n \geq 3$, then*

$$rm(G) = \begin{cases} n & \textit{if } n \textit{ is odd} \\ n+2 & \textit{if } n \textit{ is even.} \end{cases}$$

Proof Let $G = K_{1,n-1}$ where $V(G) = \{v, v_1, v_2, \ldots, v_{n-1}\}$ and $\deg v = n - 1$.
First, suppose that n is odd. Thus, $n = 2t + 1$ for some positive integer t. Define
the coloring $c : E(G) \to [n]$ by $c(vv_i) = i$ for $1 \leq i \leq t$ and $c(vv_i) = i + 1$ for
$t + 1 \leq i \leq 2t$. Since

$$cm(v) = \frac{1}{2t} \left[\sum_{i=1}^{2t+1} i - (t + 1) \right] = t + 1$$

and $cm(v_i) = c(vv_i)$ for $1 \leq i \leq 2t$, it follows that c is a rainbow mean coloring
with $rm(c) = n$. Therefore, $rm(G) = n$ if n is odd.

Next, suppose that $n \geq 4$ is even. Then $n = 2t$ for some integer $t \geq 2$. First,
we show that there is a rainbow mean coloring c of G with $rm(c) = n + 2$. Define
$c : E(G) \to \mathbb{N}$ such that

$$\{c(vv_i) : 1 \leq i \leq 2t - 1\} = [2t + 2] - \{t + 1, t + 2, 2t + 1\}.$$

Since

$$cm(v) = \frac{1}{2t - 1} \sum_{i=1}^{2t-1} c(vv_i) = \frac{1}{2t - 1} \left[\binom{2t + 3}{2} - (t + 1) - (t + 2) - (2t + 1) \right]$$

$$= \frac{1}{2t-1}[(2t+3)(t+1) - (4t+4)] = t+1$$

and $\mathrm{cm}(v_i) = c(vv_i)$ for $1 \le i \le 2t-1$, it follows that c is a rainbow mean coloring of G with $\mathrm{rm}(c) = 2t+2$. Therefore, $\mathrm{rm}(G) \le n+2$.

It remains to show that $\mathrm{rm}(G) \ge n+2 = 2t+2$. Assume, to the contrary, that there is a rainbow mean coloring c of G such that $\mathrm{rm}(c) \in \{2t, 2t+1\}$. We consider two cases, according to whether $\mathrm{rm}(c) = 2t$ or $\mathrm{rm}(c) = 2t+1$.

Case 1 $\mathrm{rm}(c) = 2t$. If t is odd, then $n \equiv 2 \pmod 4$ and all vertices of G are odd. By Corollary 5.1, no such rainbow mean coloring c exists. We show that no such rainbow mean coloring c exists regardless of the parity of t. Then $\{\mathrm{cm}(u) : u \in V(G)\} = [2t]$. Since $\mathrm{cm}(v_i) = c(vv_i)$ for $1 \le i \le 2t-1$, it follows that

$$\{c(vv_i) : 1 \le i \le 2t-1\} = [2t] - \{a\}$$

for some integer $a \in [2t]$. Thus,

$$\mathrm{cm}(v) = \frac{1}{2t-1}\left[\binom{2t+1}{2} - a\right] = \frac{1}{2t-1}[t(2t+1) - a]$$

$$= \frac{1}{2t-1}(2t^2 + t - a).$$

If $a = 1$, then $\mathrm{cm}(v) = t+1$; while if $a = 2t$, then $\mathrm{cm}(v) = t$. In either case, $\mathrm{cm}(v) = \mathrm{cm}(v_i)$ for some integer i with $1 \le i \le 2t-1$, which is impossible. On the other hand, if $1 < a < 2t$, then $\mathrm{cm}(v)$ is not an integer, which is also impossible.

Case 2 $\mathrm{rm}(c) = 2t+1$. Then $\{\mathrm{cm}(u) : u \in V(G)\} \subseteq [2t+1]$. Since $\mathrm{cm}(v_i) = c(vv_i)$ for $1 \le i \le 2t-1$, it follows that

$$\{c(vv_i) : 1 \le i \le 2t-1\} = [2t+1] - \{a, b\}$$

for some $a, b \in [2t+1]$ and $a \ne b$. Thus,

$$\mathrm{cm}(v) = \frac{1}{2t-1}\left[\binom{2t+2}{2} - (a+b)\right] = \frac{1}{2t-1}[(t+1)(2t+1) - (a+b)]$$

$$= \frac{1}{2t-1}[(2t^2 + 3t + 1) - (a+b)].$$

- If $a = 1$ and $b = 2$, then $\mathrm{cm}(v) = t+2$.
- If $a = 2t$ and $b = 2t+1$, then $\mathrm{cm}(v) = t$.
- If $\mathrm{cm}(v) = t+1$, then $a+b = 2t+2$, where $1 \le a < t+1 < b \le 2t+1$.

Fig. 5.5 A rainbow mean
coloring of $K_{3,3}$

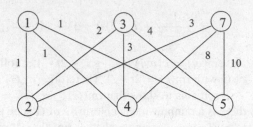

In any of these situations, $cm(v) = cm(v_i)$ for some integer i with $1 \le i \le 2t - 1$, which is impossible. For any other choice of a and b, it follows that $cm(v)$ is not an integer, which is also impossible. □

The rainbow mean index of all complete bipartite graphs that are not stars was determined in [3] using the matrix representation of edge-colored graphs technique employed in the proof of Theorem 5.4. The rainbow mean index of every complete bipartite graph is stated in the following result.

Theorem 5.6 *Let r and s be integers with $1 \le r \le s$ and $n = r + s \ge 3$. Then*

$$\text{rm}(K_{r,s}) = \begin{cases} n & \text{if } rs \text{ is even} \\ n + 1 & \text{if } rs \text{ is odd and } s \ge 3 \\ n + 2 & \text{if } s \text{ is odd and } r = 1. \end{cases}$$

For example, $\text{rm}(K_{3,3}) = 7$ by Theorem 5.6. A rainbow mean coloring c of $K_{3,3}$ with $\text{rm}(c) = 7$ is shown in Fig. 5.5.

From the results obtained on the rainbow mean index of many connected graphs G of order $n \ge 3$, the value of $\text{rm}(G)$ has always been either n or $n + 1$ with the one exception of stars of even order $n \ge 4$, which have rainbow mean index $n + 2$. In fact, the following conjecture was stated in [1].

Conjecture 5.1 For every connected graph G of order $n \ge 3$,

$$n \le \text{rm}(G) \le n + 2.$$

5.3 Bipartite Graphs

Since the only connected graphs of order $n \ge 3$ that have been shown to have rainbow mean index different from n or $n + 1$ are a single class of trees, namely stars of even order, this resulted in other classes of trees to be investigated, including double stars.

A *double star* is a tree of diameter 3. Thus, every double star T has exactly two vertices that are not leaves, which are referred to as the *central vertices* of T. For integers a and b with $2 \le a \le b$, let $S_{a,b}$ denote the double star of order $a + b$

(and size $a + b - 1$) whose central vertices have degrees a and b. The rainbow mean index of every double star was determined in [4], which is stated next.

Theorem 5.7 *For integers a and b where $a, b \geq 2$,*

$$\text{rm}(S_{a,b}) = \begin{cases} a + b & \text{if } ab \text{ is even or } ab \text{ is odd and } a + b \not\equiv 2 \pmod 4 \\ a + b + 1 & \text{if } ab \text{ is odd and } a + b \equiv 2 \pmod 4. \end{cases}$$

The rainbow mean indices of all trees that have been determined led to the following conjecture stated in [4].

Conjecture 5.2 Let T be a tree of order $n \geq 5$ that is not a star. Then $\text{rm}(T) = n$ if and only if (i) $n \not\equiv 2 \pmod 4$ or (ii) $n \equiv 2 \pmod 4$ and T has at least one even vertex; while $\text{rm}(T) = n + 1$ if $n \equiv 2 \pmod 4$ and all vertices of T have odd degrees.

Since trees form a special class of bipartite graphs, this led to the investigation of other special classes of bipartite graphs. A well-known class of bipartite graphs is that of the hypercubes. The *hypercube* Q_n is K_2 if $n = 1$, while for $n \geq 2$, Q_n is defined recursively as the Cartesian product $Q_{n-1} \square K_2$ of Q_{n-1} and K_2. For each integer $n \geq 2$, the hypercubes Q_n is an n-regular bipartite graph of order 2^n. In order to determine the rainbow mean index of every hypercube (see [3]), we first recall a well-known result (see [2]).

Theorem 5.8 *Every regular bipartite graph contains a 1-factor (in fact, is 1-factorable).*

Theorem 5.9 *For each integer $n \geq 2$, $\text{rm}(Q_n) = 2^n$.*

Proof Since the order of Q_n is 2^n, it suffices to show that there is a rainbow mean coloring c of Q_n with $\text{rm}(c) = 2^n$. We proceed by induction on $n \geq 2$. Since $\text{rm}(Q_2) = \text{rm}(C_4) = 4$ by Theorem 5.3, the statement is true for $n = 2$. Suppose that there is a rainbow mean coloring of Q_n with rainbow mean index 2^n for some integer $n \geq 2$. We show that $G = Q_{n+1} = Q_n \square K_2$ has a rainbow mean coloring c with $\text{rm}(c) = 2^{n+1}$.

Let H and H' be the two copies of Q_n in G where each vertex v in H is adjacent to the vertex v' in H'. Since Q_n is a regular bipartite graph, it follows that Q_n has a 1-factor. Let F be a 1-factor of H and let F' be the corresponding 1-factor in H'. By the induction hypothesis, there is a rainbow mean coloring $c_H : E(H) \to \mathbb{N}$ of H with $\text{rm}(c_H) = 2^n$. Thus,

$$\{\text{cm}_{c_H}(v) : v \in V(H)\} = [2^n]. \tag{5.2}$$

We now extend the coloring c_H of H to an edge coloring $c : E(G) \to \mathbb{N}$ of G by defining

$$c(e) = \begin{cases} c_H(e) & \text{if } e \in E(H) \cup [E(H') - E(F')] \\ c_H(e) + (n+1)2^n & \text{if } e \in E(F') \\ i & \text{if } e = vv' \text{ and } \text{cm}_{c_H}(v) = i \text{ for } 1 \le i \le 2^n. \end{cases}$$

It remains to show that c is a rainbow mean coloring with $\text{rm}(c) = 2^{n+1}$.

- Let $v \in V(H)$, where $\text{cm}_{c_H}(v) = i \in [2^n]$. Since

$$(n+1)\,\text{cm}_c(v) = ni + i = (n+1)i,$$

it follows that $\text{cm}(v) = \text{cm}_{c_H}(v)$. Hence, $\{\text{cm}_c(v) : v \in V(H)\} = [2^n]$ by (5.2).
- Let $v' \in V(H')$, where v is the neighbor of v' in H. Then $\text{cm}_{c_H}(v) = i$ for some integer $i \in [2^n]$. By the defining property of c, it follows that

$$(n+1)\,\text{cm}_c(v') = n\text{cm}_{c_H}(v) + i + (n+1)2^n$$

$$= ni + i + (n+1)2^n = (n+1)(i + 2^n).$$

Since $\deg_G v' = n + 1$, it follows that $\text{cm}_c(v') = i + 2^n$. Hence,

$$\{\text{cm}_c(v') : v' \in V(H')\} = [2^n + 1, 2^{n+1}].$$

This implies that c is a rainbow mean coloring of G with $\text{rm}(c) = 2^{n+1}$. Hence, by mathematical induction, $\text{rm}(Q_n) = 2^n$ for each integer $n \ge 2$. $\square$

To illustrate the proof of Theorem 5.9, we construct a rainbow mean coloring c of Q_4 with $\text{rm}(c) = 2^4 = 16$ from a rainbow mean coloring c_H of $H = Q_3$ with $\text{rm}(c_H) = 2^3 = 8$. This coloring c is shown in Fig. 5.6, where the four edges in the 1-factor F in H and the four edges in the 1-factor F' is H' are drawn in bold. Furthermore, $c(e') = c_H(e) + 4 \cdot 2^3 = c_H(e) + 32$ where $e \in E(F)$. Thus,

$$\{c(e) : e \in E(F)\} = \{1, 3, 6, 8\} \text{ and } \{c(e') : e' \in E(F')\} = \{33, 35, 38, 40\}.$$

A proof similar to that of Theorem 5.9 can be used to prove the following result.

Theorem 5.10 *If G is a connected regular bipartite graph of order $n \ge 3$ with $\text{rm}(G) = n$, then $\text{rm}(G \square K_2) = 2n$.*

There, is in fact, a more general result on Cartesian products of certain class of graphs.

Theorem 5.11 *Let G be a connected regular graph of order $n \ge 4$ with $\text{rm}(G) = n$ containing a 1-factor and let H be a connected graph of order p. Then $\text{rm}(G \square H) = np$.*

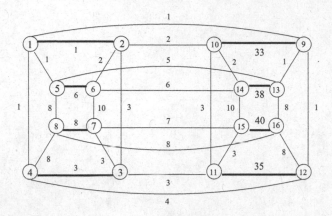

Fig. 5.6 A rainbow mean coloring of Q_4 constructed from a rainbow mean coloring of Q_3

The *prism* $C_n \square K_2$, $n \geq 3$, is the Cartesian product of the n-cycle C_n and K_2. Of course, $C_n \square K_2$ is bipartite if and only if n is even. The rainbow mean index of every prism is stated next.

Theorem 5.12 *For each integer* $n \geq 3$,

$$\text{rm}(C_n \square K_2) = \begin{cases} 2n & \text{if } n \text{ is even} \\ 2n+1 & \text{if } n \text{ is odd.} \end{cases}$$

By Conjecture 5.1 if G is a connected graph of order $n \geq 3$, then $\text{rm}(G) \leq n+2$. All connected graphs considered thus far substantiate this conjecture. As we have previously stated, the only connected graphs G of order $n \geq 3$ that we have seen with $\text{rm}(G) = n+2$ are stars of even order. Consequently, not only may this conjecture be true but those connected graphs G of order n for which $\text{rm}(G) = n+2$ may be rare.

References

1. G. Chartrand, J. Hallas, E. Salehi, P. Zhang, Rainbow mean colorings of graphs. Discrete Math. Lett. **2**, 18–25 (2019)
2. G. Chartrand, P. Zhang, *Chromatic Graph Theory*, 2nd edn. (Chapman & Hall/CRC Press, Boca Raton, 2020)
3. J. Hallas, E. Salehi, P. Zhang, Rainbow mean colorings of bipartite graphs. Bull. Inst. Combin. Appl. **88**, 78–97 (2020)
4. J. Hallas, E. Salehi, P. Zhang, Rainbow mean colorings of trees. Springer Proceedings in Mathematics & Statistics. To appear

Chapter 6
Royal Colorings

In 1985, Harary and Plantholt [7] introduced another way to obtain an irregular labeling (or weighting or coloring) of the vertices of a graph by assigning integers of the set $[k]$ to the edges of the graph. For each vertex v, rather than adding or averaging the colors of the edges incident with v, they assigned the set of colors of the edges incident with v, with the goal of minimizing k so that distinct vertices have distinct subsets of $[k]$ assigned to them. These subsets then became the colors of the vertices. Over the years, this type of coloring has gone by many names. Here, we will refer to these colorings as majestic colorings. Since the vertex colors are subsets of the set $[k]$ in a majestic coloring, this led to a more general coloring where the colors assigned to the edges are themselves subsets of $[k]$, rather than elements of $[k]$, and the color of a vertex v is the union of the colors of the edges incident with v. Once again, the goal is to minimize k so that distinct vertices have distinct colors, that is, distinct subsets of $[k]$, assigned to them. Here too, such colorings have gone by different names. In this book, we refer to these colorings as royal colorings.

6.1 The Majestic Index of a Graph

Let G be a connected graph of order 3 or more. For a positive integer k, let $\mathscr{P}^*([k])$ denote the set of nonempty subsets of $[k] = \{1, 2, \ldots, k\}$. An edge coloring

$$c : E(G) \rightarrow [k]$$

is a *majestic coloring* (or *majestic k-coloring*) if

$$c' : V(G) \rightarrow \mathscr{P}^*([k])$$

© The Author(s), under exclusive license to Springer Nature Switzerland AG 2021
A. Ali et al., *Irregularity in Graphs*, SpringerBriefs in Mathematics,
https://doi.org/10.1007/978-3-030-67993-4_6

Fig. 6.1 A graph with
majestic index 3

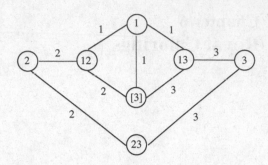

is an irregular vertex coloring for which $c'(v)$ is the set of colors of the edges
incident with v for each vertex v of G. The minimum positive integer k for which G
has a majestic k-coloring is the *majestic index* of G, denoted by maj(G). Figure 6.1
shows a majestic 3-coloring of a graph G of order 7, where, for simplicity, we write
the set $\{a\}$ as a and $\{a, b\}$ as ab, and so maj$(G) \leq 3$. Since there are only three
nonempty subsets of [2], it follows that maj$(G) > 2$. Therefore, maj$(G) = 3$.

While there is no majestic coloring of the graph K_2, every connected graph of
order 3 or more (size 2 or more) has a majestic coloring.

Theorem 6.1 *If G is a connected graph of size $m \geq 2$, then* maj$(G) \leq m$.

Proof Let $E(G) = \{e_1, e_2, \ldots, e_m\}$ where $m \geq 2$ and let $c : E(G) \to [m]$ be
defined by $c(e_i) = i$ for $1 \leq i \leq m$. Since no two vertices of G have the same set
of incident edges, it follows that $c'(u) \neq c'(v)$ for every pair u, v of distinct vertices
of G. Therefore, maj$(G) \leq m$.   □

The upper bound in Theorem 6.1 is sharp since for the star $K_{1,m}$ of size $m \geq 2$,
it follows that maj$(K_{1,m}) = m$. Harary and Plantholt [7] determined the majestic
index of all complete graphs of order at least 3.

Theorem 6.2 *For every integer $n \geq 3$,*

$$\text{maj}(K_n) = \lceil \log_2 n \rceil + 1.$$

Proof First, we show that maj$(K_n) \geq \lceil \log_2 n \rceil + 1$. Let maj$(K_n) = k$. Then there
exists a majestic k-coloring $c : E(K_n) \to [k]$. Therefore, if u and v are distinct
vertices of K_n, then $c'(u) \neq c'(v)$. Since $c(uv) \in c'(u) \cap c'(v)$, it follows that
$c'(u) \cap c'(v) \neq \emptyset$. Because $c'(u) \neq c'(v)$, one of $c'(u)$ and $c'(v)$ contains an element
$\ell \in [k]$ that the other does not have, say $c'(u)$ contains ℓ and $\ell \notin c'(v)$. Since $\ell \in$
$\overline{c'(v)}$, it follows that $\overline{c'(v)} \neq \emptyset$. If K_n contains a vertex x for which $c'(x) \subseteq \overline{c'(v)}$,
then $c'(x) \cap c'(v) = \emptyset$, which is impossible. Consequently, for each color $i \in \overline{c'(v)}$
and for each vertex x of K_n, we have $c'(x) \subseteq [k] - \{i\}$. Therefore, there are at most
2^{k-1} possible choices for $c'(x)$. Thus, $n \leq 2^{k-1}$ and so $\log_2 n \leq k - 1$. Hence,
maj$(K_n) = k \geq \lceil \log_2 n \rceil + 1$.

Next, we show that maj$(K_n) \leq \lceil \log_2 n \rceil + 1$. Here, we let $r = \lceil \log_2 n \rceil + 1$ and
show that there exists a majestic r-coloring of K_n. Since $r = \lceil \log_2 n \rceil + 1$, we have

Fig. 6.2 A majestic 3-coloring of K_3

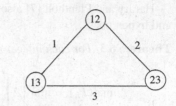

Fig. 6.3 A majestic 4-coloring of K_6

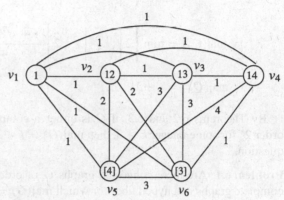

$r \geq \log_2 n + 1$ and so $n \leq 2^{r-1}$. As shown in Fig. 6.2, there is a majestic 3-coloring of K_3 and so $\text{maj}(K_n) \leq \lceil \log_2 n \rceil + 1$ when $n = 3$. We may therefore assume that $n \geq 4$ and so $n \geq r + 1$.

Let $V(K_n) = \{v_1, v_2, \ldots, v_r, \ldots, v_n\}$. We now assign distinct subsets S_i ($1 \leq i \leq n$) of the set $[r]$ containing 1 to the vertices of K_n in a manner that produces a majestic r-coloring of K_n whose resulting vertex colors are exactly the subsets of $[r]$ that were assigned to the vertices of K_n. Let $S_1 = \{1\}$. For $2 \leq i \leq r$, let $S_i = \{1, i\}$ and $S_{r+1} = [r]$. Since $n \leq 2^{r-1}$ and there are 2^{r-1} subsets of $[r]$ not containing 1, there are subsets $S_{r+2}, S_{r+3}, \ldots, S_n$ of $[r]$ containing 1 that are distinct from $S_1, S_2, \ldots, S_{r+1}$. We now assign the subset S_i to the vertex v_i for $1 \leq i \leq n$. Since $1 \in S_i$ for $1 \leq i \leq n$, it follows that $S_i \cap S_j \neq \emptyset$ for $1 \leq i < j \leq n$. Define the coloring $c : E(K_n) \to [r]$ so that $c(v_i v_j)$ is the largest color in $S_i \cap S_j$. This edge coloring c is illustrated for the complete graph K_6 in Fig. 6.3 where $r = \lceil \log_2 6 \rceil + 1 = 4$.

We now show that the edge coloring c is a majestic r-coloring of K_n. For $1 \leq i \leq n$, define $c'(v_i)$ to be the set of colors of the edges incident with v_i. Therefore, $c'(v_i) \subseteq S_i$ for $1 \leq i \leq n$. First, $c'(v_1) = \{1\} = S_1$. For $2 \leq i \leq r$, since $v_1 v_i, v_i v_{r+1} \in E(K_n)$, it follows that $c'(v_i) = \{1, i\} = S_i$. Next, we consider $c'(v_i)$ for $r + 1 \leq i \leq n$. Suppose that $\ell \in S_i$, where $\ell \neq 1$. Since ℓ is the largest color in S_ℓ and $v_i v_\ell \in E(K_n)$, it follows that $c(v_i v_\ell) = \ell$. Therefore, $\ell \in c'(v_i)$ and so $c'(v_i) = S_i$ for all i ($1 \leq i \leq n$). Since the sets S_i, $1 \leq i \leq n$, are distinct, the coloring c is a majestic r-coloring of K_n. Hence, $\text{maj}(K_n) \leq r = \lceil \log_2 n \rceil + 1$. □

Harary and Plantholt [7] also determined the majestic index of all paths, cycles, and hypercubes.

Theorem 6.3 *For each integer $n \geq 3$,*

$$\mathrm{maj}(P_n) = \min \left\{ 2 \left\lceil \sqrt{\frac{n-1}{2}} \right\rceil, \ 2 \left\lceil \frac{1 + \sqrt{8n-9}}{4} \right\rceil - 1 \right\}$$

$$\mathrm{maj}(C_n) = \min \left\{ 2 \left\lceil \sqrt{\frac{n}{2}} \right\rceil, \ 2 \left\lceil \frac{1 + \sqrt{8n+1}}{4} \right\rceil - 1 \right\}$$

$$\mathrm{maj}(Q_n) = n + 1.$$

By Theorems 6.2 and 6.3, if G is either a complete graph or a hypercube of order 2^k for some integer $k \geq 3$, then $\mathrm{maj}(G) = k + 1$. This suggests the following question.

Problem 6.1 Are there connected graphs G of order 2^k where $k \geq 3$ other than complete graphs and hypercubes for which $\mathrm{maj}(G) = k + 1$?

6.2 The Royal Index of a Graph

In a majestic coloring of a connected graph G of order 3 or more, each edge of G is assigned a color from a set $[k]$ for a positive integer k in a manner that requires the set of colors of the edges incident with a vertex of G to be distinct from the set of colors of the edges incident with every other vertex of G. By assigning the subset $\{i\} \subseteq [k]$ to an edge of G rather than the element $i \in [k]$ and defining the color of a vertex to be the union of the colors of its incident edges, this not only provides an equivalent manner to consider majestic colorings but suggests a more general coloring, potentially only requiring a set $[\ell]$ from which to select colors for an integer $\ell < k$.

For a connected graph G of order 3 or more, let $c : E(G) \rightarrow \mathscr{P}^*([k])$ be an edge coloring of G for some positive integer k. From the edge coloring c, a vertex coloring $c' : V(G) \rightarrow \mathscr{P}^*([k])$ results defined by

$$c'(v) = \bigcup_{e \in E_v} c(e),$$

where E_v is the set of edges of G incident with v. If $c'(x) \neq c'(y)$ for every pair x, y of distinct vertices of G (that is, if c' is an irregular vertex coloring of G), then c is called a *royal coloring* (or *royal k-coloring*) of G. The minimum positive integer k for which a graph G has a royal k-coloring is the *royal index* of G, denoted by $\mathrm{roy}(G)$. This concept was introduced and studied independently in [2] and [3],

Fig. 6.4 A graph G with
$\text{roy}(G) = 3$ and $\text{maj}(G) = 4$

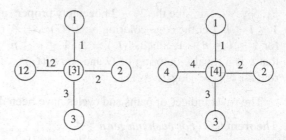

a royal 3-coloring   a majestic 4-coloring

respectively, and has been studied further in [1]. Since every majestic coloring of a connected graph G of order 3 or more is also a royal coloring of G, the following observation is immediate.

Corollary 6.1 *Every connected graph G of order 3 or more has a royal coloring and so $\text{roy}(G)$ exists. Furthermore, $\text{roy}(G) \leq \text{maj}(G)$.*

Figure 6.4 shows a royal 3-coloring and a majestic 4-coloring of the star $G = K_{1,4}$. We have seen that $\text{maj}(G) = 4$. Since there are only three nonempty sets of $[2]$, it follows that $\text{roy}(G) > 2$ and so $\text{roy}(G) = 3$.

The example $K_{1,4}$ in Fig. 6.4 shows that there are connected graphs G for which $\text{roy}(G) < \text{maj}(G)$. In fact, there are connected graphs G such that $\text{maj}(G)$ is sufficiently larger than $\text{roy}(G)$. We have seen that if $n \geq 4$, then $\text{maj}(K_{1,n-1}) = n - 1$. To determine $\text{roy}(K_{1,n-1})$, the following result is useful.

Theorem 6.4 *If G is a connected graph of order $n \geq 4$, then*

$$\text{roy}(G) \geq \lceil \log_2(n + 1) \rceil = \lfloor \log_2 n \rfloor + 1.$$

Proof Let $\text{roy}(G) = k$ and let $c : E(G) \rightarrow \mathscr{P}^*([k])$ be a royal k-coloring of G, where $c' : V(G) \rightarrow \mathscr{P}^*([k])$ is the resulting vertex coloring of G. Since $c'(v) \neq \emptyset$ for each vertex v of G and $|\mathscr{P}^*([k])| = 2^{k-1}$, it follows that $n \leq 2^{k-1}$ and so

$$\text{roy}(G) = k \geq \lceil \log_2(n + 1) \rceil = \lfloor \log_2 n \rfloor + 1,$$

as desired.   □

Theorem 6.5 *For every integer $n \geq 4$,*

$$\text{roy}(K_{1,n-1}) = \lceil \log_2(n + 1) \rceil.$$

Proof Let $G = K_{1,n-1}$, where $V(G) = \{v, v_1, v_2, \ldots, v_{n-1}\}$ such that v is the center of the star G. By Theorem 6.4, it suffices to show that $\text{roy}(K_{1,n-1}) \leq \lceil \log_2(n + 1) \rceil$. Let $k = \lceil \log_2(n + 1) \rceil$. We show that G has a royal k-coloring. Since $\lceil \log_2(n + 1) \rceil \geq 3$, it follows that $2^{k-1} - 1 \leq n - 1 \leq 2^k - 2$. Let

$S_1, S_2, \ldots, S_{2^k-2}$ be the $2^k - 2$ nonempty proper subsets of $[k]$, where $S_i = \{i\}$ for $1 \le i \le k$. Let the edge coloring $c : E(G) \to \mathscr{P}^*([k])$ be defined by $c(vv_i) = S_i$ for $1 \le i \le n - 1$. Since $c'(v_i) = S_i$, $1 \le i \le n - 1$, and $c'(v) = [k]$, it follows that c is a royal k-coloring of G and so $\mathrm{roy}(G) \le k = \lceil \log_2(n + 1) \rceil$. Therefore, $\mathrm{roy}(K_{1,n-1}) = \lceil \log_2(n + 1) \rceil$.   □

The royal indices of paths and cycles have been determined (see [1–3]).

Theorem 6.6 *For each integer $n \ge 4$,*

$$\mathrm{roy}(P_n) = \lfloor \log_2 n \rfloor + 1$$

$$\mathrm{roy}(C_n) = \begin{cases} \lfloor \log_2 n \rfloor + 2 & \text{if } n = 7 \\ \lfloor \log_2 n \rfloor + 1 & \text{if } n \ne 7. \end{cases}$$

It is easy to see that $\mathrm{roy}(C_3) = 3$. Figure 6.5 shows royal $\mathrm{roy}(C_n)$-colorings of C_n for $3 \le n \le 7$. Here again, we write the set $\{a\}$ as a and $\{a, b\}$ as ab. As Theorem 6.6 indicates, the royal index of C_7 does not follow the same pattern of $\mathrm{roy}(C_n)$ for all other values of $n \ge 4$. For this reason, we verify the value of $\mathrm{roy}(C_7)$. The royal 4-coloring of C_7 shown in Fig. 6.5 verifies that $\mathrm{roy}(C_7) \le 4$. Since the order of C_7 is $7 = 2^3 - 1$, it still must be shown that there is no royal 3-coloring of C_7. Suppose that there is a royal 3-coloring of C_7. Then all seven colors $1, 2, 3, 12, 12, 23$ and $[3]$ must be used in the resulting vertex coloring of C_7. In particular, there are three vertices of C_7 colored 1, 2, and 3. Therefore, there must be two adjacent edges of C_7 colored with each of 1, 2, and 3. However, regardless of how the seventh edge of C_7 is colored, it is impossible for the set of vertex colors to be $\mathscr{P}^*([3])$. Thus, $\mathrm{roy}(C_7) = 4$.

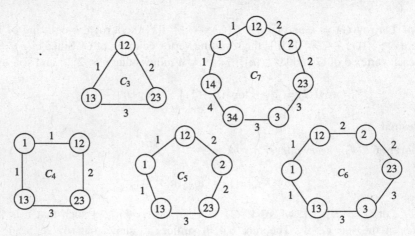

Fig. 6.5 Royal colorings of C_n where $3 \le n \le 7$

There are also familiar connected graphs G for which $\text{roy}(G) = \text{maj}(G)$, including complete graphs and hypercubes (see [1–3]).

Proposition 6.1 *For every integer* $n \geq 4$,

$$\text{roy}(K_n) = \text{maj}(K_n) = \lceil \log_2 n \rceil + 1.$$

Proof Since $\text{maj}(K_n) = \lceil \log_2 n \rceil + 1$ by Theorem 6.2 and $\text{roy}(K_n) \leq \text{maj}(K_n)$ by Corollary 6.1, it remains to show that $\text{roy}(K_n) \geq \lceil \log_2 n \rceil + 1$. Let $k = \text{roy}(K_n)$. Thus, there exists a royal k-coloring $c : E(K_n) \to \mathscr{P}^*([k])$ of K_n, which therefore results in an irregular vertex coloring $c' : V(K_n) \to \mathscr{P}^*([k])$. Let u and v be two vertices of K_n. Since $c(uv) \subseteq c'(u) \cap c'(v)$, it follows that $c'(u)$ and $c'(v)$ are distinct, nonempty, non-disjoint subsets of $[k]$. Therefore, for every subset A of $[k]$ for which $c'(x) = A$ for a vertex x of K_n, there is no vertex y of K_n for which $c'(y) = \overline{A} = [k] - A$. Thus, there are at least n subsets of $[k]$ none of which can be a color of any vertex of K_n. Hence, the number of possible colors for the n vertices of K_n is at most $2^k - n$ and so $n \leq 2^k - n$. Therefore, $n \leq 2^{k-1}$ and so $\text{roy}(K_n) = k \geq \lceil \log_2 n \rceil + 1$. Consequently, $\text{roy}(K_n) = \lceil \log_2 n \rceil + 1$. $\qquad\square$

Theorem 6.7 *For every integer* $n \geq 3$,

$$\text{roy}(Q_n) = \text{maj}(Q_n) = n + 1.$$

Proof By Theorem 6.3, $\text{maj}(Q_n) = n + 1$ and by Corollary 6.1, $\text{roy}(Q_n) \leq \text{maj}(Q_n)$. Thus, it remains to show that $\text{roy}(Q_n) \geq n + 1$. Since the order of Q_n is 2^n, it follows by Theorem 6.4 that $\text{roy}(Q_n) \geq \lceil \log_2(2^n + 1) \rceil = n + 1$. Therefore, $\text{roy}(Q_n) = n + 1$. $\qquad\square$

We have seen that the royal indices of graphs belonging to several familiar classes have been determined, but determining the royal index of graphs in general appears to be a challenging task. For this reason, research in this area has turned to obtaining bounds for the royal index. One such upper bound for the royal index of a graph G is expressed in terms of the royal indices of the spanning trees of G (see [2, 3]).

Theorem 6.8 *If G is a connected graph of order 4 or more, then*

$$\text{roy}(G) \leq 1 + \min\{\text{roy}(T) : T \text{ is a spanning tree of } G\}. \tag{6.1}$$

Proof Among all spanning trees of G, let T be one having the minimum royal index, say $\text{roy}(T) = k$. Then there exists a royal k-coloring $c_T : E(T) \to \mathscr{P}^*([k])$ of T. Then $c'_T(x) \neq c'_T(y)$ for every two vertices x and y of T. The royal k-coloring c_T is then extended to a royal $(k+1)$-coloring $c_G : E(G) \to \mathscr{P}^*([k+1])$ of G by defining

$$c_G(e) = \begin{cases} c_T(e) & \text{if } e \in E(T) \\ \{k+1\} & \text{if } e \in E(G) - E(T). \end{cases}$$

If a vertex v of G is incident only with edges of T, then $c'_G(v) = c'_T(v)$, while if v is incident with one or more edges of G not belonging to T, then $c'_G(v) = c'_T(v) \cup \{k + 1\}$. Since c'_T is an irregular vertex coloring of T, it follows that c'_G is an irregular vertex coloring of G. Hence, c_G is a royal $(k + 1)$-coloring of G. Consequently, $\text{roy}(G) \le k + 1 = \text{roy}(T) + 1$ and so (6.1) holds.   □

The upper bound given in Theorem 6.8 for the royal index of a connected graph shows the value of knowing the royal index of trees. We have seen that the royal indices of all stars and paths were determined in [2, 3] and that of other classes of trees were obtained in [1]. In every instance, it has been shown that $\text{roy}(T) = \lfloor \log_2 n \rfloor + 1$ for every tree T of order $n \ge 4$ belonging to any of these classes. This led to the following conjecture stated in [3].

Conjecture 6.1 If T is a tree of order $n \ge 4$, then

$$\text{roy}(T) = \lfloor \log_2 n \rfloor + 1.$$

We saw in Theorem 6.4 that if G is a connected graph of order $n \ge 4$, then

$$\text{roy}(G) \ge \lfloor \log_2 n \rfloor + 1.$$

Since there is a unique integer $k \ge 3$ such that $2^{k-1} \le n \le 2^k - 1$, it follows that $\text{roy}(G) \ge k$. If Conjecture 6.1 is true, then $\text{roy}(T) = k$ for every tree T of order $n \ge 4$ with $2^{k-1} \le n \le 2^k - 1$. Furthermore, by Theorem 6.8, if Conjecture 6.1 is true, then $\text{roy}(G) \le k + 1$ for every connected graph G of order $n \ge 4$ with $2^{k-1} \le n \le 2^k - 1$. This leads to a conjecture about the royal indices of connected graphs in general, also stated in [3].

Conjecture 6.2 If G is a connected graph of order $n \ge 4$ where $2^{k-1} \le n \le 2^k - 1$ for an integer $k \ge 3$, then $\text{roy}(G) = k$ or $\text{roy}(G) = k + 1$.

If Conjecture 6.2 is true, it then follows that every connected graph G of order $n \ge 4$ where $2^{k-1} \le n \le 2^k - 1$ possesses a royal $(k + 1)$-coloring. We have seen that the royal index of every tree of order n with $2^{k-1} \le n \le 2^k - 1$ is either k or conjectured to be k and that every cycle of order n with $2^{k-1} \le n \le 2^k - 1$ other than C_7 has royal index k. This suggests that every connected graph G of order n with $2^{k-1} \le n \le 2^k - 1$ and small size is likely to have royal index k. For connected graphs of large size, the following result was obtained in [1].

Theorem 6.9 *Let G be a graph of order $n \ge 4$ and size m where $2^{k-1} \le n \le 2^k - 1$ for some integer $k \ge 3$. If $m > \frac{1}{2}(4^k - 3^k - 2^k + 1)$, then $\text{roy}(G) \ge k + 1$.*

Proof First, we construct a graph of order $2^k - 1 = |\mathscr{P}^*([k])|$, which we denote by G_k. The vertices of G_k are labeled with the $2^k - 1$ distinct elements of $\mathscr{P}^*([k])$. The label of each vertex v of G_k is denoted by $\ell(v)$. Consequently,

$$\{\ell(v) : v \in V(G_k)\} = \mathscr{P}^*([k]).$$

Fig. 6.6 The graph G_3

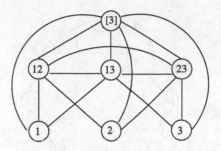

For $1 \le i \le k$, let

$$V_i = \{v \in V(G_k) : |\ell(v)| = i\}.$$

Therefore, $|V_i| = \binom{k}{i}$ for $1 \le i \le k$ and $\{V_1, V_2, \dots, V_k\}$ is a partition of $\mathscr{P}^*([k])$. Two vertices u and v of G_k are adjacent in G_k if and only if $\ell(u) \cup \ell(v) \ne \emptyset$. Let m_k denote the size of G_k. The graph G_3 of order $7 = 2^3 - 1$ and size $m_3 = 15$ is shown in Fig. 6.6. Observe that if G is not a subgraph of the graph G_k, then $\mathrm{roy}(G) \ge k+1$.

Let $v \in V_i$ for some i with $1 \le i \le k$. Thus, $\ell(v) = S$ is an i-element subset of $[k]$. There are $2^i - 1$ nonempty subsets of S and 2^{k-i} subsets of $[k] - S$. For each nonempty subset S' of S and each subset T of $[k] - S$, the vertex v is adjacent to that vertex w of G_k for which $\ell(w) - S' \cup T$. Since v is not adjacent to itself, however, it follows that

$$\deg_{G_k} v = (2^i - 1)2^{k-i} - 1.$$

Furthermore, there are $\binom{k}{i}$ vertices in V_i for $1 \le i \le k$. Therefore, the size of G_k is

$$m_k = \frac{1}{2} \sum_{i=1}^{k} \binom{k}{i} \left[(2^i - 1)2^{k-i} - 1 \right] = \frac{1}{2} \sum_{i=1}^{k} \binom{k}{i} (2^k - 2^{k-i} - 1)$$

$$= \frac{1}{2} \left[\sum_{i=1}^{k} \binom{k}{i} 2^k - \sum_{i=1}^{k} \binom{k}{i} 2^{k-i} - \sum_{i=1}^{k} \binom{k}{i} \right]$$

$$= \frac{1}{2} \left[2^k \sum_{i=1}^{k} \binom{k}{i} - 2^k \sum_{i=1}^{k} \binom{k}{i} \left(\frac{1}{2}\right)^i - \sum_{i=1}^{k} \binom{k}{i} \right]$$

$$= \frac{1}{2} \left\{ 2^k(2^k - 1) - 2^k \left[\left(1 + \frac{1}{2}\right)^k - 1 \right] - (2^k - 1) \right\}$$

$$= \frac{1}{2}(4^k - 3^k - 2^k + 1).$$

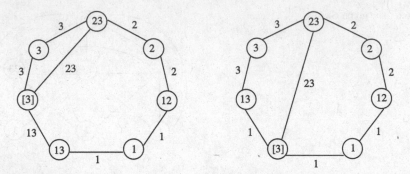

Fig. 6.7 Showing that roy$(C_7 + e) = 3$ for each $e \notin E(C_7)$

Since G is a graph of order $n \geq 4$ and size m where $2^{k-1} \leq n \leq 2^k - 1$ such that

$$m > m_k = \frac{1}{2}(4^k - 3^k - 2^k + 1),$$

then G is not a subgraph of G_k and consequently roy$(G) \geq k + 1$.   □

By the proof of Theorem 6.9 if G is a connected graph of order $n \geq 4$ and size m where $2^{k-1} \leq n \leq 2^k - 1$ such that $m > m_k$, then $G \nsubseteq G_k$ and consequently roy$(G) \geq k + 1$. In fact, if G possesses any property that implies that $G \nsubseteq G_k$, then roy$(G) \geq k + 1$. For example, if the order of G is $n = 2^k - 1$ and $\delta(G) \geq \delta(G_k) + 1$ or G has more than one vertex of degree $n - 1$, then roy$(G) \geq k + 1$. On the other hand, even though $C_7 \subseteq G_3$ (where $n = 2^3 - 1$ and $k = 3$), $|E(C_7)| = 7 < m_3$, and $\delta(C_7) < \delta(G_3)$, we saw that roy$(C_7) = 4 = k + 1$. Furthermore, for every chord e of C_7, roy$(C_7 + e) = 3$ (see Fig. 6.7). Therefore, while one might suspect that roy$(G + uv) \geq$ roy(G) for every connected graph G and every pair u, v of nonadjacent vertices of G, such is not the case.

What we have seen is that if G is a connected graph of order $n \geq 4$ where $2^{k-1} \leq n \leq 2^k - 1$ having a sufficiently large size, then roy$(G) \neq k$. However, if G is a connected graph of order $n \geq 4$ where $2^{k-1} \leq n \leq 2^k - 1$ having a small size, then we are not guaranteed that roy$(G) = k$. Indeed, even the royal indices of trees is in doubt.

References

1. A. Ali, G. Chartrand, J. Hallas, P. Zhang, Extremal problems in royal colorings of graphs. J. Combin. Math. Combin. Comput. (2019). Preprint
2. N. Bousquet, A. Dailly, E. Duchêne, H. Kheddouci, A. Parreau, A Vizing-like theorem for union vertex-distinguishing edge coloring. Discrete Appl. Math. **232**, 88–98 (2017)
3. G. Chartrand, J. Hallas, P. Zhang, Royal colorings in graphs. Ars Combin. (2019). Preprint

4. G. Chartrand, J. Hallas, P. Zhang, Color-induced graph colorings. J. Combin. Math. Combin. Comput. (2015)
5. G. Chartrand, P. Zhang, *Chromatic Graph Theory*, 2nd edn. (Chapman & Hall/CRC Press, Boca Raton, 2020)
6. E. Györi, M. Horňák, C. Palmer, M. Woźniak, General neighbour-distinguishing index of a graph. Discrete Math. **308**, 827–831 (2008)
7. F. Harary and M. Plantholt, The point-distinguishing chromatic index. *Graphs and Applications*. Wiley, New York (1985), 147–162.
8. M. Horňák, R. Sotók, General neighbour-distinguishing index via chromatic number. Discrete Math. **310**, 1733–1736 (2010)

Chapter 7
Traversable Irregularity

An Eulerian circuit in a connected graph G is a circuit that contains every edge of G exactly once while an Eulerian walk in G is a closed walk that contains every edge of G at least once. While only Eulerian graphs contain an Eulerian circuit, every nontrivial connected graph contains an Eulerian walk. The irregularity concept here is an irregular Eulerian walk in G, which is an Eulerian walk where no two edges of G are encountered the same number of times. The irregularity counterpart for vertices is a Hamiltonian walk. These are the primary topics for this chapter.

7.1 Introduction

The origin of graph theory is often attributed to Leonhard Euler's solution of the famous Königsberg bridge problem in 1736. During that period, the city of Königsberg was the capital of German East Prussia. The River Pregel flowed through Königsberg separating the city into four land regions. Seven bridges were built over the river. Figure 7.1 shows a map of Königsberg with its four land regions (labeled A, B, C, D), the location of the river and the seven bridges (labeled a, b, c, d, e, f, g). The story goes that during the 1730s, some citizens of Königsberg enjoyed strolling about the city and some wondered whether it was possible to walk about the city and pass over each of its seven bridges exactly once. This problem became known as the *Königsberg bridge problem*.

Leonhard Euler, the great Swiss mathematician, became aware of this problem. Although he did not think this problem was particularly mathematical in nature, he solved the problem. In fact, not only did he show that it was impossible to walk about Königsberg crossing each bridge exactly once, he presented a generalization of the problem and solved this problem as well (although his solution lacked some important details). The paper by Euler [6] containing all of this information was published in 1736. In this connection, Ore [12] made the following statement:

A. Ali et al., *Irregularity in Graphs*, SpringerBriefs in Mathematics,
https://doi.org/10.1007/978-3-030-67993-4_7

Fig. 7.1 A map of
Königsberg

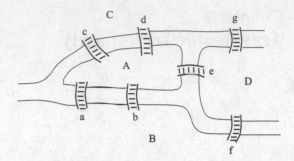

Fig. 7.2 A multigraph
representing Königsberg

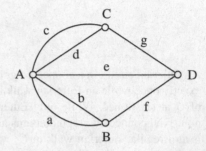

The theory of graphs is one of the few fields of mathematics with a definite birth date.

While the term "graph" never appeared in Euler's paper, Euler's reasoning was clearly graph theoretic in nature. (Indeed, the term "graph," as used in graph theory, was not introduced until 1878, when the British mathematician James Joseph Sylvester first used this word.)

Expressed in terms of graphs, Euler's paper contained the following result:

> Suppose that there is a town consisting of land regions, some pairs of which are joined by one or more bridges. For a town possessing such a round-trip, the vertices of the corresponding graph or multigraph G are the land regions where every two vertices of G are joined by a number of edges equal to the number of bridges joining the corresponding land regions. There is a round-trip in the town passing over each bridge exactly once if and only if G is connected and every vertex of G has even degree.

Such a graph or multigraph is referred to as Eulerian. Since each vertex in the multigraph representing the map of Königsberg has odd degree (as shown in Fig. 7.2), it follows by Euler's result that it was not possible to walk about Königsberg and pass over each bridge exactly once. Euler's paper not only solved the Königsberg bridge problem, it marked the beginning of graph theory.

7.2 The Chinese Postman Problem

If every edge of a nontrivial connected graph G of size m is replaced by two edges, then the resulting multigraph is Eulerian, which implies that G contains a closed walk in which every edge of G appears exactly twice. Euler made this observation in his paper. Of course, G contains a closed walk in which every edge of G appears exactly once if and only if G itself is Eulerian. A weighted graph H can be obtained from G by assigning a positive integer weight to each edge of G. The degree of a vertex v in H is the sum of the weights of the edges incident with v. From the observation above, every vertex of H is even if every edge of G is assigned the weight 2. If every edge of G is assigned the weight 1, then every vertex of H is even if and only if G is Eulerian. Thus, it is always possible to assign each edge of G a weight 1 or 2 in such a way that every vertex in the resulting weighted graph is even.

A problem of interest is that of determining the minimum sum of all positive integer weights assigned to the edges of G so that every vertex in the resulting weighted graph is even. This is equivalent to determining the minimum length of a closed walk in G that contains every edge of G at least once. A solution to this problem also provides a solution to the so-called Chinese Postman Problem, named by Alan Goldman for the Chinese mathematician Meigu Guan (often known as Mei-Ko Kwan) who introduced this problem in 1962. Suppose that a postman starts from the post office and has mail to deliver to the houses along each street on his mail route. Once he has completed delivering the mail, he returns to the post office. The problem is to find the minimum length of a round trip that accomplishes this, as we state next.

The Chinese Postman Problem *Determine the minimum length of a round trip that traverses every road in a mail route at least once.*

The minimum length of a closed walk that contains every edge of a connected graph G of size m at least once is m if G is Eulerian. If G is not Eulerian, then G contains $2k$ odd vertices for some positive integer k. Suppose that the $2k$ odd vertices are divided into k pairs and the distance between the vertices in each pair is determined and these k numbers are summed. If d is the minimum value of all such sums over all partitions of these $2k$ odd vertices into k pairs, then the minimum length of such a closed walk is $m+d$. Suppose that $\{\{u_1, v_1\}, \{u_2, v_2\}, \ldots, \{u_k, v_k\}\}$ is a partition of the $2k$ odd vertices such that $\sum_{i=1}^{k} d(u_i, v_i) = d$ and P_i is a $u_i - v_i$ path of length $d(u_i, v_i)$ for $i = 1, 2, \ldots, k$. Then the paths P_i are pairwise edge-disjoint. If we replace each edge that belongs to one of these paths by two parallel edges, then we obtain an Eulerian multigraph M of size $m + d$. An Eulerian circuit in M gives rise then to a closed walk of minimum length in G that contains every edge of G at least once—a solution to the Chinese Postman Problem.

We now describe some of the terminology and notation we will need for this chapter, some of which we have introduced earlier but will be useful to repeat. Here, we only consider nontrivial connected graphs.

For vertices u and v in a graph G, a $u - v$ *walk* W in G is a sequence

$$W = (u = v_0, v_1, v_2, \ldots, v_k = v) \tag{7.1}$$

of vertices in G such that $v_{i-1}v_i$ is an edge of G for each i ($1 \le i \le k$). If $e_i = v_{i-1}v_i$, then the walk W in (7.1) can also be denoted by

$$W = (e_1, e_2, \ldots, e_k). \tag{7.2}$$

The length of the walk W is denoted by $L(W)$ and so $L(W) = k$ for the walk W in (7.1) and (7.2).

If G is a multigraph rather than a graph, then some pairs of vertices are joined by more than one edge. In this case, it is necessary to denote a walk as a sequence of edges as in (7.2) rather than a sequence of vertices as in (7.1) to avoid confusion. If $u = v$, then the $u - v$ walk is *closed*; while if $u \ne v$, then the $u - v$ walk is *open*. If there is no repetition of edges in a walk, then the walk is a *trail*. A closed nontrivial trail is a *circuit*. A $u - v$ walk W as in (7.1) is a $u - v$ *path* if the vertices $v_0, v_1, \ldots, v_k$ are distinct. If W is a circuit for which the vertices $v_0, v_1, \ldots, v_{k-1}$ are distinct, then W is a *cycle*.

A circuit in a graph G that contains every edge of G is an *Eulerian circuit*, while an open trail containing every edge of G is an *Eulerian trail*. A graph containing an Eulerian circuit is an *Eulerian graph* and a graph containing an Eulerian trail is a *traversable graph*. In 1736, Leonhard Euler established the following characterization of Eulerian graphs [6].

Theorem 7.1 (Euler's Theorem) *A nontrivial connected graph G is Eulerian if and only if every vertex of G has even degree.*

An important corollary of Theorem 7.1 is the following characterization of traversable graphs.

Corollary 7.1 *A nontrivial connected graph G is traversable if and only if G contains exactly two vertices of odd degree. Any Eulerian trail in G then begins at one of these vertices and terminates at the other.*

If a graph G has four or more odd vertices, then G contains neither an Eulerian circuit nor an Eulerian trail, which again explains why there was no journey about Königsberg that crosses each bridge exactly once. Even though these graphs do not contain Eulerian circuits or Eulerian trails, there are some interesting properties that these graphs possess. We saw in the discussion related to finding a solution to the Chinese Postman Problem that if a connected graph G contains $2k$ odd vertices, then these $2k$ odd vertices can be partitioned into k pairs resulting in k pairwise edge-disjoint paths in G, each connecting pairs of odd vertices. In fact, it is well known that G itself can be decomposed to k open trails connecting odd vertices.

Theorem 7.2 *If G is a connected graph containing $2k \geq 4$ odd vertices, then G can be decomposed into k open trails connecting odd vertices but no fewer.*

7.3 Irregular Eulerian Walks

While the Chinese Postman Problem asks for the minimum length of a closed walk in a connected graph G such that every edge of G appears on the walk once or twice, another problem of interest is that of determining the minimum length of a closed walk in G in which no two edges of G appear the same number of times. Such walks in a graph G distinguish the edges of G by their occurrences on the walk, which gives rise to the concept of irregular Eulerian walks in graphs.

Let G be a nontrivial connected graph of size m. By an *Eulerian walk* in G, we mean a closed walk that contains every edge of G. Thus, the length of an Eulerian walk W in G is m if and only if W is an Eulerian circuit. In general then, the minimum length of an Eulerian walk in G is $m + d$ for some nonnegative integer d. We saw that if every edge of G is replaced by two parallel edges, then the resulting multigraph M is Eulerian and each Eulerian circuit in M gives rise to an Eulerian walk in G that encounters every edge of G exactly twice. Hence, if G is not Eulerian, then the minimum length of an Eulerian walk in G is more than m but not more than $2m$ and every edge appears once or twice in such an Eulerian walk in G. This concept was studied in [7, 10].

Let H be a weighted graph obtained by assigning weights (positive integers) to the edges of a connected graph G. Recall that the *degree* $\deg_H v$ of a vertex v in H is the sum of the weights of the edges incident with v. Determining the minimum length of an Eulerian walk in G is then equivalent to determining an assignment of the weights 1 or 2 to the edges of G such that the sum of these weights is minimum and the degree of every vertex in H is even. The subgraph induced by the edges labeled 2 is the union of edge-disjoint paths in G. As we mentioned in Sect. 7.2, this problem is directly related to a well-known problem called the Chinese Postman Problem, which is the problem of determining the minimum length of a round trip that traverses every road in a mail route at least once.

For every nontrivial connected graph G of size m, there is always an Eulerian walk in which each edge of G is encountered the same number of times. An *irregular Eulerian walk* in G is an Eulerian walk that encounters no two edges of G the same number of times. Thus, the length of an irregular Eulerian walk in G is at least

$$1 + 2 + \cdots + m = \binom{m+1}{2}.$$

If $E(G) = \{e_1, e_2, \ldots, e_m\}$ and each edge e_i $(1 \leq i \leq m)$ of G is replaced by $2i$ parallel edges, then the resulting multigraph M is Eulerian and each Eulerian circuit in M gives rise to an irregular Eulerian walk in which each edge e_i of G appears exactly $2i$ times in the walk. Thus, G contains an irregular Eulerian walk of length

$$2 + 4 + 6 + \cdots + 2m = 2\binom{m+1}{2} = m^2 + m.$$

The length of a walk W is denoted by $L(W)$. If W is an irregular Eulerian walk of minimum length in a connected graph G of size m, then

$$\binom{m+1}{2} \le L(W) \le 2\binom{m+1}{2}.$$

The concept of irregular Eulerian walks in graphs was first introduced and studied in [1] and studied further in [2–4]. A problem of interest here is that of determining the minimum length of an irregular Eulerian walk in G, which we refer to as the *Eulerian irregularity* of G, which is denoted by $EI(G)$. Therefore, if G is a connected graph of size m, then

$$\binom{m+1}{2} \le EI(G) \le 2\binom{m+1}{2}. \tag{7.3}$$

Both bounds in (7.3) are sharp. First, we show that the lower bound in (7.3) is sharp. For an odd integer $n \ge 5$, let $G = C_n^2$ be the square of the n-cycle C_n. That is, if $C_n = (v_1, v_2, \ldots, v_n, v_1)$, then

$$E(G) = \{v_1 v_2, v_2 v_3, \ldots, v_{n-1} v_n, v_n v_1\} \cup$$

$$\{v_1 v_3, v_3 v_5, \ldots, v_{n-2} v_n, v_n v_2, v_2 v_4, \ldots, v_{n-3} v_{n-1}, v_{n-1} v_1\}.$$

Thus, G is a 4-regular graph of size $m = 2n$ and

$$C = (v_1, v_2, v_3, \ldots, v_n, v_1, v_3, v_5, \ldots, v_n, v_2, v_4, \ldots, v_{n-3}, v_{n-1}, v_1)$$

is an Eulerian circuit of G. Suppose that C encounters the edges $e_1, e_2, \ldots, e_m$ in this order and each edge e_i $(1 \le i \le m)$ is replaced by i parallel edges. Then the resulting multigraph M is Eulerian and so each Eulerian circuit in M gives rise to an irregular Eulerian walk in which each edge e_i of G appears exactly i times in the walk. Thus, $EI(G) = \binom{m+1}{2}$ and so the lower bound in (7.3) is sharp (see [1]). To see that the upper bound in (7.3) is sharp, we first state a theorem due to Kwan [11].

Kwan's Theorem *Let G be a connected graph and let W be a closed walk of minimum length containing every edge of G at least once. Then W encounters no edge of G more than twice and no more than half of the edges in any cycle appear twice.*

Theorem 7.3 *For a connected graph G of size $m \geq 1$,*

$$EI(G) = 2\binom{m+1}{2} \text{ if and only if } G \text{ is a tree.}$$

Proof First, we show that if uv is a bridge of G, then uv occurs an even number of times on an Eulerian walk of G. Let W be an Eulerian walk of G with initial vertex u. Then the first time that v is encountered on W, it is preceded by u and the next time that u is encountered on W, it is preceded by v. Therefore, uv occurs an even number of times on W.

If G is a tree, then every edge of G is a bridge and so each edge of G is encountered an even number of times on W. Therefore, $EI(G) \geq 2\binom{m+1}{2}$. It then follows by (7.3) that $EI(G) = 2\binom{m+1}{2}$. For the converse, assume that G is not a tree. Then G contains at least one cycle. By Kwan's theorem, there is an Eulerian walk W in which no edge of G occurs on W more than twice and some edges occur on W exactly once. Let $e_1, e_2, \ldots, e_{k'} (k \geq 1)$ be those edges occurring exactly once on W and let $f_1, f_2, \ldots, f_\ell$ be those edges occurring exactly twice on W. By assigning each edge e_i ($1 \leq i \leq k$) the weight $2i - 1$ and each edge f_j ($1 \leq j \leq \ell$) the weight $2j$ if $\ell \geq 1$, we obtain a weighted graph in which every vertex is even. Thus, there is an Eulerian walk in G where e_i appears $2i - 1$ times and f_j appears $2j$ times. Since there is an irregular Eulerian walk of length less than $2\binom{m+1}{2}$, it follows that $EI(G) < 2\binom{m+1}{2}$.   □

If we were to consider all sets S of m positive integers and label the edges of a connected graph G of size m with distinct elements of S in such a way that every vertex is even in the resulting weighted graph, then the minimum of the sums of the elements in all such sets S is $EI(G)$.

7.4 Optimal Irregular Eulerian Walks

An irregular Eulerian walk W in a connected graph G of size m is said to be *optimal* if $L(W) = \binom{m+1}{2}$. In this case, the edges of G can be ordered as $e_1, e_2, \ldots, e_m$ such that e_i ($1 \leq i \leq m$) is encountered exactly i times in W. As we have seen, there are graphs that possess an optimal irregular Eulerian walk. First, we present a characterization of such connected graphs (see [1]).

Theorem 7.4 *Let G be a connected graph of size m. Then G contains an optimal irregular Eulerian walk if and only if G contains a subgraph of size $\lceil m/2 \rceil$, every vertex of which is even.*

Proof First, assume that G contains a subgraph F of size $\lceil m/2 \rceil$ such that every vertex of F is even. Then

$$E(G) = \{e_1, e_2, \ldots, e_{\lceil m/2 \rceil}\} \cup \{e_1', e_2', \ldots, e_{\lfloor m/2 \rfloor}'\},$$

where $E(F) = \{e_1, e_2, \ldots, e_{\lceil m/2 \rceil}\}$. We construct an Eulerian multigraph M by replacing each edge e_i where $1 \le i \le \lceil m/2 \rceil$ by $2i - 1$ parallel edges and replacing each edge e'_j, where $1 \le j \le \lfloor m/2 \rfloor$, by $2j$ parallel edges. An Eulerian circuit in M gives rise to an irregular Eulerian walk W in G such that each edge e_i of G appears exactly $2i - 1$ times in W, where $1 \le i \le \lceil m/2 \rceil$, and each edge e'_j of G appears exactly $2j$ times in W where $1 \le j \le \lfloor m/2 \rfloor$. Then the length of W is

$$1 + 2 + 3 + \cdots + m = \binom{m+1}{2}$$

and W is an optimal irregular Eulerian walk in G.

For the converse, suppose that G contains an optimal irregular Eulerian walk W. We may assume that $E(G) = \{f_1, f_2, \ldots, f_m\}$, where f_i appears exactly i times $(1 \le i \le m)$ on W. Let F be the subgraph of G of size of $\lceil m/2 \rceil$ induced by the set

$$\{f_1, f_3, \ldots, f_{2\lfloor (m-1)/2 \rfloor + 1}\}$$

and let F' be the subgraph of G induced by the set

$$\{f_2, f_4, \ldots, f_{2\lceil (m-1)/2 \rceil}\}.$$

Thus, $\{F, F'\}$ is a decomposition of G. We claim that every vertex of F is even. Let M be the weighted graph obtained by assigning the weight i $(1 \le i \le m)$ to each edge f_i of G. Let H be the weighted subgraph of M induced by the edges of F and let H' be the weighted subgraph of M induced by the edges of F'. Since G has an Eulerian walk in which each edge f_i appears exactly i times, every vertex of M has even degree. Since $\deg_M v = \deg_H v + \deg_{H'} v$ for every vertex v of G and $\deg_M v$ and $\deg_{H'}$ are both even, it follows that $\deg_H v$ is even. Suppose that $\deg_F v = k$. Then v is incident with k edges in H, each of odd weight. Since $\deg_H v$ is even, k is even and so v is an even vertex in F.   □

By Theorem 7.4, the graphs G_1 and G_3 of Fig. 7.3 contain optimal irregular Eulerian walks while G_2 and G_4 do not. Since the Petersen graph P has size 15 and P contains an 8-cycle, it follows by Theorem 7.4 that P contains an optimal irregular Eulerian walk. On the other hand, by Theorem 7.4, no cycle contains an optimal irregular Eulerian walk. In fact, for each integer $m \ge 3$,

$$EI(C_m) = 1 + 3 + 5 + \cdots + (2m - 1) = m^2.$$

We have seen that if $n \ge 5$ is odd, then C_n^2 contains an optimal irregular Eulerian walk. Since the size of C_n^2 is $2n$ and C_n is a 2-regular subgraph of size n in C_n^2, it follows by Theorem 7.4 that C_n^2 contains an optimal irregular Eulerian walk for each integer $n \ge 4$.

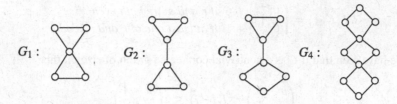

Fig. 7.3 Illustrating Theorem 7.4

By Theorem 7.4, if G is a connected graph of size m, then $EI(G) = \binom{m+1}{2}$ if and only if G contains a subgraph of size $\lceil m/2 \rceil$, every vertex of which is even. The following is also a consequence of Theorem 7.4.

Corollary 7.2 *If G is a connected bipartite graph of size $m \geq 1$ such that $m \equiv 1$ (mod 4) or $m \equiv 2$ (mod 4), then G does not contain an optimal irregular Eulerian walk.*

Proof If $m \equiv 1$ (mod 4) or $m \equiv 2$ (mod 4), then $\lceil m/2 \rceil$ is odd. If G contains an optimal irregular Eulerian walk, then by Theorem 7.4, G contains a subgraph H of size $\lceil m/2 \rceil$, each of whose vertices is even. Therefore, H is a bipartite graph of odd size, each vertex of which is even. This is impossible.   □

All complete graphs and complete bipartite graphs containing an optimal irregular Eulerian walk have been determined in [1].

Theorem 7.5 *For each integer $n \geq 2$, the complete graph K_n contains an optimal irregular Eulerian walk if and only if $n \geq 4$.*

Theorem 7.6 *For integers r and s with $2 \leq r \leq s$, the complete bipartite graph $K_{r,s}$ contains an optimal irregular Eulerian walk if and only if*

 (i) r and s are both even and $(r, s) \neq (2, 4k + 2)$ for any nonnegative integer k or
(ii) at least one of r and s is odd and $rs \not\equiv 1, 2$ (mod 4).

The Eulerian irregularity of every complete graph and every complete bipartite graph has been determined in [3].

Theorem 7.7 *For each integer $n \geq 2$,*

$$EI(K_n) = \begin{cases} 2 & \text{if } n = 2 \\ 9 & \text{if } n = 3 \\ \binom{\binom{n}{2}+1}{2} & \text{if } n \geq 4. \end{cases}$$

Theorem 7.8 *If the complete bipartite graph $K_{r,s}$ is not optimal where $2 \leq r \leq s$, then*

$$EI(K_{r,s}) = \begin{cases} \binom{rs+1}{2} + 6 & \text{if } r \text{ and } s \text{ are both even} \\ \binom{rs+1}{2} + 1 & \text{if at least one of } r \text{ and } s \text{ is odd.} \end{cases}$$

We have seen that if G is a nontrivial connected graph of size m, then

$$\binom{m+1}{2} \leq EI(G) \leq 2\binom{m+1}{2}.$$

This gives rise to the following question:

> For given positive integers k and m with $\binom{m+1}{2} \leq k \leq 2\binom{m+1}{2}$, is there a connected graph G of size m such that $EI(G) = k$?

In [3] a necessary and sufficient condition was established for all pairs k, m of positive integers for which there is a nontrivial connected graph G of size m with $EI(G) = k$.

Theorem 7.9 *Let k and m be positive integers with $\binom{m+1}{2} \leq k \leq 2\binom{m+1}{2}$. Then there exists a nontrivial connected graph G of size m with $EI(G) = k$ if and only if there exists an integer x with $0 \leq x \leq m$ and $x \neq 1, 2$ such that $x^2 + (m - x)(m - x + 1) = k$.*

7.5 Irregular Hamiltonian Walks

A *Hamiltonian walk* in a connected graph G is a spanning walk in G. Thus, if W is a Hamiltonian walk in a connected graph G of order n, then the order of W (the number of vertices encountered, including multiplicities) is $n + d$ for some nonnegative integer d. Consequently, the order of W is n if and only if W is a Hamiltonian path of G. This concept has been studied in [5, 8, 9]. A Hamiltonian walk W of a graph G is *irregular* if no two vertices of G are encountered the same number of times in W. The following two observations are useful.

Observation 7.10 *The order of an irregular Hamiltonian walk in a connected graph of order n is at least $\sum_{i=1}^{n} i = \binom{n+1}{2}$.*

Observation 7.11 *Let G be a connected graph and T a spanning tree of G. If T contains an irregular Hamiltonian walk, then so does G.*

Theorem 7.12 *Every connected graph contains an irregular Hamiltonian walk.*

Proof By Observation 7.11, it suffices to verify the statement for trees. It is easy to observe that the statement is true for every tree of order n where $1 \leq n \leq 4$. This is illustrated for the two trees T_1 and T_2 of order 4 of Fig. 7.4, where $W_1 = (u, v, w, x, w, v, w, x, w, x)$ is an irregular Hamiltonian walk in T_1 and $W_2 = (u, v, w, v, w, v, x, v, x, v, x)$ is an irregular Hamiltonian walk in T_2.

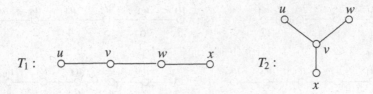

Fig. 7.4 Irregular Hamiltonian walks in trees of order 4

Assume that every tree of order n contains an irregular Hamiltonian walk where $n \geq 4$. Let T be a tree of order $n + 1$ and let u be an end-vertex of T where v is the neighbor of u in T. Since $T' = T - u$ is a tree of order n, it follows by the induction hypothesis that T' contains an irregular Hamiltonian walk W'. Suppose that the maximum number of times that a vertex of T' is encountered in W' is k and the vertex v is encountered ℓ times in W' where then $\ell \leq k$. At a place in W' where v is encountered, we insert the walk u, v a total of $k + 1$ times immediately after v, producing a Hamiltonian walk W in T. Thus, u is encountered $k + 1$ times in W, v is encountered $\ell + k + 1 \geq k + 2$ times in W, and all other vertices of T are encountered the same number of times on W as they were encountered in W'. Hence, W is an irregular Hamiltonian walk of T.   □

The *Hamiltonian irregularity index* of a connected graph G, denoted by $HI(G)$, is the minimum order of an irregular Hamiltonian walk in G. If G has order n, then

$$HI(G) \geq \sum_{i=1}^{n} i = \binom{n+1}{2}.$$

An irregular Hamiltonian walk of order $\binom{n+1}{2}$ in a connected graph G of order n is called an *optimal irregular Hamiltonian walk* in G. A question of interest here is the following:

Which connected graphs contain an optimal irregular Hamiltonian walk?

The following observation is useful.

Observation 7.13 *For every connected graph G,*

$$HI(G) \leq \min\{HI(T) : T \text{ a spanning tree of } G\}.$$

Observation 7.13 indicates the value of knowing the Hamiltonian irregularity indices of trees, especially those trees possessing an optimal irregular Hamiltonian walk. In addition to the paths P_1, P_2, and P_3, the four paths P_4, P_5, P_6, and P_7 also have an optimal irregular Hamiltonian walk. To see this, let $P_n = (u_1, u_2, \ldots, u_n)$ for $4 \leq n \leq 7$. We define a walk W_n in P_n ($4 \leq n \leq 7$) as follows:

$$W_4 = (u_1, u_2, u_3, u_4, u_3, u_2, u_3, u_2, u_3, u_4)$$

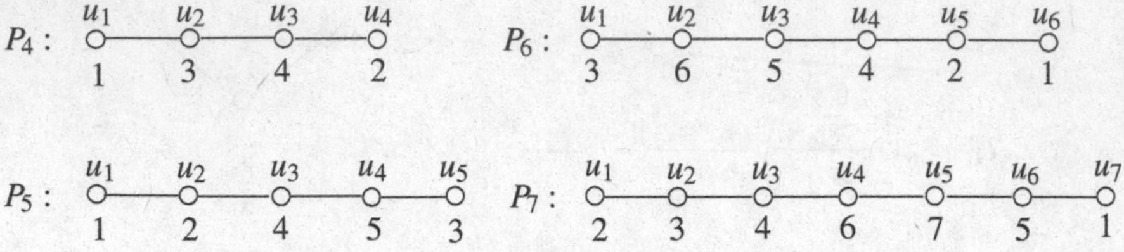

P_4 :
$$\begin{array}{cccc} u_1 & u_2 & u_3 & u_4 \\ \circ\!\!-\!\!\!-\!\!\!-\!\!\!-\!\!\circ\!\!-\!\!\!-\!\!\!-\!\!\!-\!\!\circ\!\!-\!\!\!-\!\!\!-\!\!\!-\!\!\circ \\ 1 & 3 & 4 & 2 \end{array}$$

P_6 :
$$\begin{array}{cccccc} u_1 & u_2 & u_3 & u_4 & u_5 & u_6 \\ \circ\!\!-\!\!\circ\!\!-\!\!\circ\!\!-\!\!\circ\!\!-\!\!\circ\!\!-\!\!\circ \\ 3 & 6 & 5 & 4 & 2 & 1 \end{array}$$

P_5 :
$$\begin{array}{ccccc} u_1 & u_2 & u_3 & u_4 & u_5 \\ \circ\!\!-\!\!\circ\!\!-\!\!\circ\!\!-\!\!\circ\!\!-\!\!\circ \\ 1 & 2 & 4 & 5 & 3 \end{array}$$

P_7 :
$$\begin{array}{ccccccc} u_1 & u_2 & u_3 & u_4 & u_5 & u_6 & u_7 \\ \circ\!\!-\!\!\circ\!\!-\!\!\circ\!\!-\!\!\circ\!\!-\!\!\circ\!\!-\!\!\circ\!\!-\!\!\circ \\ 2 & 3 & 4 & 6 & 7 & 5 & 1 \end{array}$$

Fig. 7.5 Showing that W_n is an optimal irregular Hamiltonian walk in P_n for $4 \leq n \leq 7$

$$W_5 = (u_1, u_2, u_3, u_4, u_5, u_4, u_3, u_2, u_3, u_4, u_5, u_4, u_3, u_4, u_5)$$

$$W_6 = (u_2, u_1, u_2, u_3, u_4, u_3, u_2, u_1, u_2, u_3, u_4, u_5, u_4, u_3,$$

$$u_2, u_1, u_2, u_3, u_4, u_5, u_6)$$

$$W_7 = (u_2, u_1, u_2, u_1, u_2, u_3, u_4, u_3, u_4, u_3, u_4, u_3, u_4, u_5, u_4, u_5, u_4,$$

$$u_5, u_6, u_5, u_6, u_5, u_6, u_5, u_6, u_5, u_6, u_7).$$

For each integer n with $4 \leq n \leq 7$, the walk W_n is an optimal irregular Hamiltonian walk in P_n, as is shown in Fig. 7.5, where the integer listed below each vertex in a path indicates the number of times this vertex is encountered in the walk given. Observe that the terminal vertex of the walk W_n in P_n is u_n for $4 \leq n \leq 7$.

Theorem 7.14 *Every path contains an optimal irregular Hamiltonian walk.*

Proof We have already seen that the paths P_n, where $1 \leq n \leq 7$, have an optimal irregular Hamiltonian walk. We may assume that $n \geq 8$. Therefore, $n = r + 4k$ for integers $r \in \{4, 5, 6, 7\}$ and $k \geq 1$. Suppose first that $k = 1$. We saw that there is an optimal irregular Hamiltonian walk W_0 in $P_r = (u_1, u_2, \ldots, u_r)$ for $4 \leq r \leq 7$ with terminal vertex u_r. We now show that there is an optimal irregular Hamiltonian walk in

$$P_{r+4} = (u_1, u_2, \ldots, u_r, u_{r+2}, u_{r+4}, u_{r+3}, u_{r+1})$$

for $4 \leq r \leq 7$ with terminal vertex u_{r+1}. Let W_1 be the walk of P_{r+4} obtained by following the walk W_0 with

(1) u_{r+2}, u_{r+4} a total of $r + 2$ times,
(2) u_{r+3}, u_{r+4} twice, and
(3) u_{r+3}, u_{r+1} a total of $r + 1$ times.

Thus, the vertex u_{r+i}, $1 \leq i \leq 4$, occurs $r + i$ times in W_1 and all other vertices of P_{r+4} are encountered the same number of times on W_1 as they were encountered in W_0. Thus, W_1 is an optimal irregular Hamiltonian walk of P_{r+4}. If $k \geq 2$, then this procedure is continued $k - 1$ times by beginning with W_1 having terminal vertex u_{r+1}. For example, if $k = 2$, let $s = r + 4$ and let $P_s = (v_1, v_2, \ldots, v_s)$. As we saw, P_s has an optimal irregular Hamiltonian walk W_1 with terminal vertex v_s. Let

$$P_{s+4} = (v_1, v_2, \ldots, v_s, v_{s+2}, v_{s+4}, v_{s+3}, v_{s+1}).$$

Let W_2 be the walk of P_{s+4} obtained by following the walk W_1 with

(1) v_{s+2}, v_{s+4} a total of $s + 2$ times,
(2) v_{s+3}, v_{s+4} twice, and
(3) v_{s+3}, v_{s+1} a total of $s + 1$ times.

Thus, the vertex v_{s+i}, $1 \le i \le 4$, occurs $s + i$ times in W_2 and all other vertices of P_{s+4} are encountered the same number of times on W_2 as they were encountered in W_1. Thus, W_2 is an optimal irregular Hamiltonian walk of P_{s+4} having terminal vertex v_{s+1}. $\square$

There are trees different from paths that have an optimal irregular Hamiltonian walk. For example,

$$W_1 = (u_1, u_2, u_3, u_2, u_3, u_2, u_3, u_2, u_3, u_2,$$
$$u_4, u_5, u_4, u_5, u_4, u_5, u_4, u_6, u_4, u_6, u_4) \text{ and}$$

$$W_2 = (u_2, u_1, u_2, u_1, u_2, u_1, u_2, u_1, u_2, u_3,$$
$$u_4, u_3, u_4, u_3, u_4, u_3, u_5, u_3, u_5, u_3, u_6)$$

are optimal irregular Hamiltonian walks in the trees T_1 and T_2 shown in Fig. 7.6, where the integer listed near each vertex in a tree indicates the number of times this vertex is encountered in the given walk.

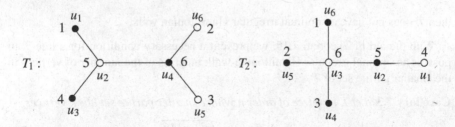

Fig. 7.6 Two optimal trees T_1 and T_2

88   7 Traversable Irregularity

The following theorem gives a necessary condition for a tree to possess an optimal irregular Hamiltonian walk.

Theorem 7.15 *Let T be a tree of order n whose smaller partite set has k vertices. If*

$$\sum_{i=1}^{k}(n+1-i) < \left\lfloor \frac{1}{2}\binom{n+1}{2}\right\rfloor,$$

then T does not have an optimal irregular Hamiltonian walk.

Proof Let $V(T) = \{v_1, v_2, \ldots, v_n\}$ and let V_1 and V_2 be partite sets of T where $k = |V_1| \le |V_2| = n - k$. Suppose that T has an optimal irregular Hamiltonian walk W where v_i is encountered $\ell(v_i)$ times on W where $1 \le i \le n$. Thus,

$$\sum_{i=1}^{n}\ell(v_i) = \sum_{i=1}^{n}i = \binom{n+1}{2}.$$

Since the vertices on W alternate between V_1 and V_2, either

(1) $\displaystyle\sum_{v\in V_1}\ell(v) = \sum_{v\in V_2}\ell(v) = \frac{1}{2}\binom{n+1}{2}$ which can only occur if $n \equiv 0, 3 \pmod 4$
    or

(2) one of $\displaystyle\sum_{v\in V_1}\ell(v)$ and $\displaystyle\sum_{v\in V_2}\ell(v)$ is 1 greater than the other, that is, one is $\left\lfloor \frac{1}{2}\binom{n+1}{2}\right\rfloor$

and the other is $\left\lceil \frac{1}{2}\binom{n+1}{2}\right\rceil$ which can only occur if $n \equiv 1, 2 \pmod 4$.

Consequently, if

$$\sum_{i=1}^{k}(n+1-i) < \left\lfloor \frac{1}{2}\binom{n+1}{2}\right\rfloor,$$

then T does not have an optimal irregular Hamiltonian walk.   □

With the aid of Theorem 7.15, we present a necessary condition for a tree T to possess an optimal irregular Hamiltonian walk in terms of the number of vertices in the smaller partite set of T.

Corollary 7.3 *Let T be a tree of order n whose smaller partite set has k vertices.*

- *If $n \equiv 0, 3 \pmod 4$ and $k \le \left\lfloor \frac{(2n+1)-\sqrt{n^2+(n+1)^2}}{2}\right\rfloor$, then T does not have an optimal irregular Hamiltonian walk.*

- *If $n \equiv 1, 2 \pmod 4$ and $k \le \left\lfloor \frac{(2n+1)-\sqrt{n^2+(n+1)^2+4}}{2}\right\rfloor$, then T does not have an optimal irregular Hamiltonian walk.*

The converse of Corollary 7.3 is not true, however. To see this, we first present an observation.

Fig. 7.7 A tree T of order 8
that does not have an optimal
irregular Hamiltonian walk

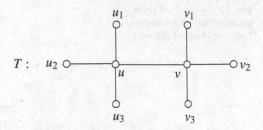

Observation 7.16 *Let T be a tree of order n and let W be an irregular Hamiltonian walk in T. For $u \in V(T)$, let $\ell(u)$ be the number of times u is encountered on W. If $v \in V(T)$ and $v_1, v_2, \ldots, v_d$ are end-vertices of T that are neighbors of v, then*

$$\ell(v) \geq \sum_{i=1}^{d} \ell(v_i).$$

The tree T in Fig. 7.7 has order $n = 8$ and each partite set of T contains four vertices. Thus,

$$4 = k > \left\lfloor \frac{(2n+1) - \sqrt{n^2 + (n+1)^2}}{2} \right\rfloor = \left\lfloor \frac{17 - \sqrt{64 + 81}}{2} \right\rfloor = 2.$$

However, T does not have an optimal irregular Hamiltonian walk. Assume, to the contrary, that T has an optimal irregular Hamiltonian walk W.

For each $x \in V(T)$, let $\ell(x)$ be the number of times x is encountered on W. Thus, $\{\ell(x) : x \in V(T)\} = [8]$. We may assume that $\ell(u) < \ell(v) \leq 8$. By Observation 7.16, it follows that

$$\ell(u) \geq \sum_{i=1}^{3} \ell(u_i) \geq 1 + 2 + 3 = 6.$$

Thus, either $\ell(u) = 6$ or $\ell(u) = 7$.

- If $\sum_{i=1}^{3} \ell(u_i) = 6$, then $\ell(v) \geq \sum_{i=1}^{3} \ell(v_i) \geq 4 + 5 + 6 = 15$, a contradiction.
- If $\sum_{i=1}^{3} \ell(u_i) = 7$, then $\ell(v) \geq \sum_{i=1}^{3} \ell(v_i) \geq 3 + 5 + 6 = 14$, a contradiction.

Thus, T does not have an optimal irregular Hamiltonian walk and so the converse of Corollary 7.3 is not true in general for trees T of order n with $n \equiv 0 \pmod 4$.

Fig. 7.8 A tree T of order 7 that does not have an optimal irregular Hamiltonian walk

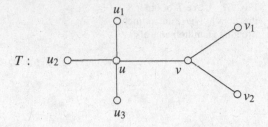

The tree T in Fig. 7.8 has order $n = 7$ and the smaller partite set of T contains three vertices. Thus,

$$3 = k > \left\lfloor \frac{(2n+1) - \sqrt{n^2 + (n+1)^2}}{2} \right\rfloor = \left\lfloor \frac{15 - \sqrt{49+64}}{2} \right\rfloor = 2.$$

However, T does not have an optimal irregular Hamiltonian walk. Assume, to the contrary, that T has an optimal irregular Hamiltonian walk W.

For each $x \in V(T)$, let $\ell(x)$ be the number of times x is encountered on W. Thus, $\{\ell(x) : x \in V(T)\} = [7]$. By Observation 7.16, it follows that

$$\ell(u) \geq \sum_{i=1}^{3} \ell(u_i) \geq 6.$$

Thus, either $\ell(u) = 6$ or $\ell(u) = 7$.

- If $\sum_{i=1}^{3} \ell(u_i) = 6$, then $\ell(v) \geq \ell(v_1) + \ell(v_2) \geq 9$, a contradiction.

- If $\sum_{i=1}^{3} \ell(u_i) = 7$, then $\ell(u) = 7$ and $\ell(v) \geq \ell(v_1) + \ell(v_2) \geq 8$, a contradiction.

Thus, T does not have an optimal irregular Hamiltonian walk and so the converse of Corollary 7.3 is not true in general for trees T of order n with $n \equiv 3 \pmod 4$. In fact, $HI(T) = 30 = \binom{8}{2} + 2$ and the walk

$$W_0 = (u_1, u, u_2, u, u_2, u, u_3, u, u_3, u, u_3, u, v, v_1, v, v_1, v, v_1, v, v_1,$$

$$v, v_2, v, v_2, v, v_2, v, v_2, v, v_2)$$

is an irregular Hamiltonian walk of order 30 in T.

References

1. E. Andrews, G. Chartrand, C. Lumduanhom, P. Zhang, On Eulerian walks in graphs. Bull. Inst. Combin. Appl. **68**, 12–26 (2013)
2. E. Andrews, C. Lumduanhom, P. Zhang, On irregular Eulerian Walks in circulants. Congr. Numer. **217**, 33–52 (2013)
3. E. Andrews, C. Lumduanhom, P. Zhang, On Eulerian irregularity in graphs. Discuss. Math. Graph Theory. **34**, 391–408 (2014)
4. E. Andrews, C. Lumduanhom, P. Zhang, On Eulerian irregularities of prisms, grids and powers of cycles. J. Combin. Math. Combin. Comput. **90**, 167–184 (2014)
5. G. Chartrand, T. Thomas, V. Saenpholphat, P. Zhang, A new look at Hamiltonian walks. Bull. Inst. Combin. Appl. **42**, 37–52 (2004)
6. L. Euler, Solutio problematis ad geometriam situs pertinentis. Comment. Acad. Sci. I. Petropolitanae **8**, 128–140 (1736)
7. S. E. Goodman, S. T. Hedetniemi, Eulerian walks in graphs. SIAM J. Comput. **2**, 16–27 (1973)
8. S.E. Goodman, S.T. Hedetniemi, On Hamiltonian walks in graphs. Congr. Numer. 335–342 (1973)
9. S.E. Goodman, S.T. Hedetniemi, On Hamiltonian walks in graphs. SIAM J. Comput. **3**, 214–221 (1974)
10. S.T. Hedetniemi, On minimum walks in graphs. Naval Res. Logist. Q. **15**, 453–458 (1968)
11. M.K. Kwan, Graphic programming using odd or even points. Acta Math. Sin. **10**, 264–266 (1960) (Chinese); translated as Chin. Math. **1**, 273–277 (1960)
12. O. Ore, *Graphs and Their Uses* (Random House, New York, 1963)

Chapter 8
Ascending Subgraph Decompositions

Among the many results in graph theory dealing with graph decompositions are those involving complete graphs. While the primary emphasis has been graph decompositions in which every two subgraphs are the same (isomorphic), here we consider graph decompositions in which not only are every two subgraphs non-isomorphic, but they possess some prescribed characteristics. Since the complete graph K_{n+1} has size $\binom{n+1}{2}$ and $\sum_{i=1}^{n} i = \binom{n+1}{2}$, this has led to decompositions, not only of K_{n+1} but of any graph G of size $\binom{n+1}{2}$ into subgraphs $G_1, G_2, \ldots, G_n$ of G such that G_i has size i for $1 \le i \le n$ with the added property that G_i is isomorphic to a subgraph of G_{i+1} for $1 \le i \le n - 1$. Such "irregular decompositions" are called ascending subgraph decompositions. It has been conjectured that every graph of size $\binom{n+1}{2}$ has such a decomposition, which is the primary topic of this chapter. A concept dealing with irregular decompositions in Ramsey theory is also discussed.

8.1 Isomorphic Decompositions

As we mentioned in Chap. 1, there has been great interest in decomposing graphs, especially complete graphs, into isomorphic subgraphs. If each subgraph is a spanning subgraph of the primary graph, then these are *isomorphic factorizations* of the graph. Examples of this mentioned in Chap. 1 are the following:

1. Every complete graph of even order is 1-factorable.
2. Every r-regular graph where $r \ge 2$ is even is 2-factorable.
3. Every complete graph of odd order 3 or more is decomposable into Hamiltonian cycles.

For graphs H and G, the graph G is *H-decomposable* if G can be decomposed into subgraphs, each isomorphic to H. Wilson [18] proved the following.

Theorem 8.1 *For every graph H without isolated vertices, there exist infinitely many positive integers n such that K_n is H-decomposable.*

In 1964, Ringel [15] introduced a conjecture which remains unresolved.

Ringel's Conjecture *If T is a tree of size m, then K_{2m+1} is T-decomposable.*

This conjecture is related to a graph labeling that has been of interest to many researchers. A nonempty graph G of size m is *graceful* if the vertices of G can be labeled with distinct elements from the set $[0, m] = \{0, 1, \ldots, m\}$ in such a way that the resulting edge labeling that assigns the positive integer $|i - j|$ to the edge joining vertices labeled i and j has all edges labeled differently. That is, the resulting edge labeling is "irregular". Necessarily then, the set of all edge labels is $[m] = \{1, 2, \ldots, m\}$. Such a labeling is called a *graceful labeling*. The connection between graceful labelings and Ringel's conjecture is the following result of Rosa [17].

Theorem 8.2 *If H is a graceful graph of size m, then K_{2m+1} is H-decomposable.*

Proof Let H be a graceful graph of size m. Then the vertices of H can be labeled with distinct elements of the set $[0, m]$ in such a way that the set of resulting edge labels is $[m]$. We now consider the complete graph K_{2m+1} with vertex set $\{v_0, v_1, \ldots, v_{2m}\}$. Place the vertices of K_{2m+1} cyclically, say clockwise, about a regular $(2m + 1)$-gon. For each integer $i \in [m]$ in the graceful labeling of H, the vertex labeled i is placed at v_i in K_{2m+1}. For each pair $i, j \in [m]$ where there is an edge of H joining vertices with these labels, an edge is drawn between v_i and v_j and this edge is labeled $|i - j|$. This results in a drawing of the graph H. Observe that $|i - j|$ is the distance between v_i and v_j on the Hamiltonian cycle $(v_0, v_1, \ldots, v_{2m}, v_0)$ of K_{2m+1}. By rotating this drawing of H through an angle of $2\pi i/(2m + 1)$ radians for $i = 1, 2, \ldots, 2m$ clockwise about the center of the $(2m + 1)$-gon, a total of $2m$ additional pairwise edge-disjoint copies of H are produced, thereby showing that K_{2m+1} is H-decomposable. $\square$

As a consequence of Theorem 8.2, if T is a graceful tree of size m, then K_{2m+1} is T-decomposable. For example, for the tree T of size 4 shown in Fig. 8.1, the

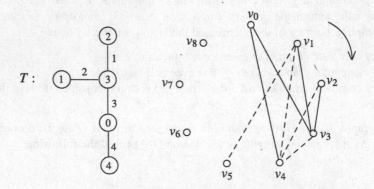

Fig. 8.1 The graph K_9 is T-decomposable

complete graph K_9 is T-decomposable where the ith copy of T ($1 \le i \le 8$) is obtained by rotating the first copy of T clockwise through an angle of $2\pi i/9$ radians.

While some graphs are graceful and many are not, no tree has been found that is not graceful. This led Anton Kotzig and Gerhard Ringel (see [12]) to make the following conjecture which, if true, establishes the truth of Ringel's Conjecture.

The Graceful Tree Conjecture *Every nontrivial tree is graceful.*

8.2 Irregular Decompositions

Proceeding along the theme of irregularity in graphs and the concept of isomorphic decompositions of graphs, where a graph is decomposed into subgraphs every two of which are isomorphic, we are led to consider the irregular version of these concepts, that is, decomposing a graph into subgraphs no two of which are isomorphic. Such a decomposition of a graph is called an *irregular decomposition* of the graph. Since every graph of size 3 or more can be decomposed into two non-isomorphic subgraphs, the primary question concerns decomposing graphs of a given size into as many non-isomorphic subgraphs (without isolated vertices) as possible, no two of which are isomorphic. For example, the complete graph $H = K_6$ of size 15 in Fig. 8.2 has an irregular decomposition into the five subgraphs $H_1 = K_3$, $H_2 = P_4$, $H_3 = 3K_2$, $H_4 = K_{1,3}$, and $H_5 = P_2 + P_3$, all of size 3, which are, in fact, the five non-isomorphic graphs of size 3 containing no isolated vertices.

The graph G of size 10 in Fig. 8.3 has an irregular decomposition into four non-isomorphic subgraphs G_1, G_2, G_3, G_4, also shown in Fig. 8.3, such that $G_1 = P_2$, $G_2 = P_3$, $G_3 = P_4$, and $G_4 = P_2 + P_4$. These subgraphs have the property that G_i is isomorphic to a subgraph of G_{i+1} for $i = 1, 2, 3$, a characteristic of interest in this chapter.

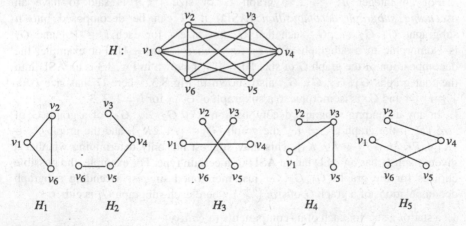

Fig. 8.2 An irregular decomposition of a graph

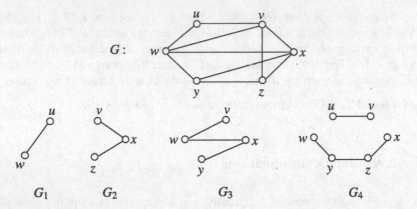

Fig. 8.3 An irregular decomposition of a graph

Because of the special interest in isomorphic decompositions of complete graphs, this suggests decomposing complete graphs into non-isomorphic subgraphs. Since the size of the complete graph K_n is $\binom{n}{2} = \frac{n(n-1)}{2}$ and $\sum_{i=1}^{n-1} i = \binom{n}{2}$, any decomposition of K_n into $n - 1$ subgraph G_i of size i for each $i \in [n - 1]$ will necessarily be an irregular decomposition of K_n. Rather than decomposing a complete graph into $n - 1$ non-isomorphic subgraphs, it is more practical to consider decomposing a complete graph into n non-isomorphic subgraphs, which suggests considering the complete graph K_{n+1} rather than K_n.

Although an irregular decomposition of K_{n+1} can be obtained in numerous ways by producing pairwise edge-disjoint subgraphs $G_1, G_2, \ldots, G_n$ without isolated vertices such that G_i has size i for each $i \in [n]$, it is of greater interest to determine whether such decompositions exist in which the subgraphs possess some specified property. This led to special irregular decompositions of graphs introduced by Alavi et al. in [1].

For an integer $n \geq 2$, a graph G of size $\binom{n+1}{2}$ is said to have an *ascending subgraph decomposition* (ASD) if G can be decomposed into n subgraphs $G_1, G_2, \ldots, G_n$ such that G_i has size i for each $i \in [n]$ and G_i is isomorphic to a subgraph of G_{i+1} for each $i \in [n - 1]$. For example, the decomposition of the graph G of size $10 = \binom{4+1}{2}$ shown in Fig. 8.3 is an ASD into the four graphs G_1, G_2, G_3, G_4, also shown in Fig. 8.3, where G_i has size i for $1 \leq i \leq 4$ and G_i is isomorphic to a subgraph of G_{i+1} for $i = 1, 2, 3$.

In any ascending subgraph decomposition $\{G_1, G_2, \ldots, G_n\}$ of a graph G of size $\binom{n+1}{2}$, the graph $G_1 = K_2$, the graph $G_2 \in \{P_3, 2K_2\}$, and the graph $G_3 \in \{K_{1,3}, P_4, 3K_2, P_3 + K_2, K_3\}$. This may suggest not only determining whether a given graph G has an ASD but an ASD of a certain type. For example, the possible choices for the graphs G_1, G_2, G_3 just mentioned suggest ascending subgraph decompositions of a graph G of size $\binom{n+1}{2}$ where each subgraph G_i is either

1. a star or a star for each of its components (a *galaxy*)
2. a path or a path for each of its components (a *linear forest*) or

3. a matching, that is, $G_i = i K_2$ for $1 \leq i \leq n$.

The most common example of a graph G of size $\binom{n+1}{2}$ is the complete graph K_{n+1}. In fact, K_{n+1} has an ASD into stars and an ASD into paths, as we show next.

Proposition 8.1 *Every complete graph K_{n+1} has an ASD into stars.*

Proof Let $V(K_{n+1}) = \{v_0, v_1, \ldots, v_n\}$ and let G_n be the subgraph of size n in K_{n+1} induced by the n edges incident with v_n. Thus, G_n is a star of size n with central vertex v_n. Deleting v_n from K_{n+1} results in the subgraph K_n of order n and size $\binom{n}{2}$. Next, let G_{n-1} be the subgraph of K_n induced by the $n-1$ edges incident with v_{n-1}. Thus, G_{n-1} is a star of size $n-1$ with central vertex v_{n-1}. Continuing in this manner, we arrive at an irregular decomposition $\{G_1, G_2, \ldots, G_n\}$ of K_{n+1} where G_i is a star of size i with central vertex v_i for $i = 1, 2, \ldots, n$. This decomposition is an ASD of K_{n+1} into stars $\square$

Proposition 8.2 *Every complete graph K_{n+1} has an ASD into paths.*

Proof If $n + 1$ is odd, then K_{n+1} can be decomposed into $n/2$ Hamiltonian cycles $H_1, H_2, \ldots, H_{n/2}$, each of size $n + 1$. We can then decompose H_1 into the paths P_2 and P_{n+1}, the cycle H_2 into the paths P_3 and P_n, and so on, finally, obtaining an irregular decomposition of K_{n+1} into the paths $P_2, P_3, \ldots, P_{n+1}$ of sizes $1, 2, \ldots, n$, respectively.

If $n + 1$ is even, then K_{n+2} can be decomposed into $(n + 1)/2$ Hamiltonian cycles $H_1, H_2, \ldots, H_{(n+1)/2}$, each of size $n + 2$. Let

$$V(K_{n+2}) = \{v_1, v_2, \ldots, v_{n+2}\}.$$

For each integer i with $1 \leq i \leq (n + 1)/2$, let

$$F_i = H_i - v_{n+2} \cong P_{n+1}.$$

Then $F_1, F_2, \ldots, F_{(n+1)/2}$ are $(n + 1)/2$ Hamiltonian paths of size n in K_{n+1}. For each integer j with $2 \leq j \leq (n + 1)/2$, we can then decompose F_j into the paths P_j and P_{n+2-j}. In particular, F_2 is decomposed into the paths P_2 and P_n, F_3 into the paths P_3 and P_{n-1} and so on $F_{(n+1)/2}$ is decomposed into $P_{(n+1)/2}$ and $P_{(n+3)/2}$. This produces an irregular decomposition of K_{n+1} into the paths $P_2, P_3, \ldots, P_{n+1} = F_1$ of sizes $1, 2, \ldots, n$. $\square$

If H is a 1-regular graph and so $E(H)$ is a matching, we refer to H itself as a *matching*. While the complete graph K_{n+1} does not have an ASD into matchings, each path and cycle of size $\binom{n+1}{2}$ does have an ASD into matchings. For example, the path P_{11} of size $10 = \binom{4+1}{2}$ can be decomposed into four matchings G_1, G_2, G_3, G_4 where $G_i = i K_2$ for $i = 1, 2, 3, 4$, as illustrated in Fig. 8.4 where each edge in G_i is labeled i. Also illustrated in Fig. 8.4 is an ASD of the cycle C_{10} of size 10

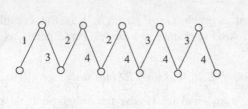
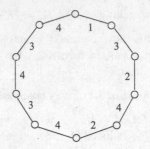

Fig. 8.4 Decomposing P_{11} and C_{10} into matchings

into four matchings H_1, H_2, H_3, H_4 such that $H_i = i K_2$ for $i = 1, 2, 3, 4$, where here as well each edge in H_i is labeled i.

The following result was observed in [1].

Proposition 8.3 *Each path and cycle of size $\binom{n+1}{2}$ has an ASD into matchings.*

8.3   The Ascending Subgraph Decomposition Conjecture

In the preceding section, many examples of graphs of size $\binom{n+1}{2}$ were given that can be decomposed into n subgraphs $G_1, G_2, \ldots, G_n$ where not only G_i has size i for each $i \in [n]$ but G_i is isomorphic to a subgraph of G_{i+1} for each $i \in [n-1]$. That is, each of these graphs of size $\binom{n+1}{2}$ has an ASD. Such ascending subgraph decompositions led to a conjecture (see [1]).

The Ascending Subgraph Decomposition Conjecture *For each integer $n \geq 2$, every graph of size $\binom{n+1}{2}$ has an ascending subgraph decomposition.*

Curiously, the very same day that Paul Erdős was informed of the ASD Conjecture (and then contributed to the initial paper on this topic) and was scheduled to give a lecture on graph theory at Western Michigan University, he mentioned this conjecture at the end of his talk and offered \$5 to anyone who could prove or disprove it.

We saw in Proposition 8.1 that every complete graph K_{n+1} has an ASD into stars. There is a related class of graphs of size $\binom{n+1}{2}$ having an ASD into stars (see [8]).

Theorem 8.3 *If G is a graph of size $\binom{n+1}{2}$, $n \geq 1$, obtained by deleting $n+1$ edges from K_{n+2}, then G has an ASD into stars.*

Proof We proceed by induction on n. The statement is obviously true for $n = 1, 2, 3$. Assume for an integer $n \geq 3$ and for any given set X of n edges in K_{n+1} that the graph $K_{n+1} - X$ has an ASD into $n - 1$ stars. We show for any given set Y of $n + 1$ edges in K_{n+2} that the graph $G = K_{n+2} - Y$ has an ASD into n stars. Let v be a vertex of G such that $\deg_G v = \Delta(G)$. Thus, either $\deg_G v = n$ or $\deg_G v = n+1$. We consider these two possibilities.

Case 1 $\deg_G v = n$. Let $e \in Y$ be the only edge in Y that is incident with v and let $X = Y - \{e\}$. Then

$$G - v \cong K_{n+1} - X.$$

By the induction hypothesis, $G - v$ has an ASD into $n - 1$ stars $K_{1,1}, K_{1,2}, \ldots,$ $K_{1,n-1}$. Let $K_{1,n}$ be the star induced by the n edges of G incident with v. This produces an ASD of G into n stars $K_{1,1}, K_{1,2}, \ldots, K_{1,n}$.

Case 2 $\deg_G v = n + 1$. Thus, no edge incident with v belongs to Y. Let $f \in Y$ and let $X = Y - \{f\}$. Then

$$H = (G - v) + f \cong K_{n+1} - X.$$

By the induction hypothesis, H has an ASD into $n - 1$ stars $K_{1,1}, K_{1,2}, \ldots, K_{1,n-1}$. Furthermore, f is an edge of some star $K_{1,i}$ ($1 \le i \le n - 1$) whose central vertex is u, say. Let E_v be the set of the $n + 1$ edges in G incident with v. Replacing f in $K_{1,i}$ by the edge uv, we obtain an ASD of the graph

$$G - E_v + uv = [(K_{n+2} - Y) - v] + uv$$

into the $n - 1$ stars $S_1 \cong K_{1,1}, S_2 \cong K_{1,2}, \ldots, S_{n-1} \cong K_{1,n-1}$. Let S_n be the star $K_{1,n}$ induced by $E_v - \{uv\}$. Then $\{S_1, S_2, \ldots, S_n\}$ is an ASD of G into n stars. □

We mentioned that a graph G is called a *galaxy* if every component of G is a star. It was shown in [8] that every galaxy of size $\binom{n+1}{2}$ has an ASD.

Theorem 8.4 *If G is a galaxy of size $\binom{n+1}{2}$, then G has an ASD.*

Necessarily, each subgraph in an ASD of a galaxy is itself a galaxy. There is a problem related to Theorem 8.4.

Which galaxies G of size $\binom{n+1}{2}$ have the property that for each subgalaxy G' of size m in G where $1 \le m \le n$, there is an ASD of G containing G'?

By Theorem 8.4, every galaxy has an ASD into galaxies. The following was stated in [8]:

Surprisingly, [this result] is the most difficult to prove. This could indicate that the ASD conjecture (if true) is a difficult one to prove.

In fact, the following problem was also stated in [8].

Problem 8.1 Let G be a graph of size $\binom{n+1}{2}$. Does G have an ASD into galaxies?

In Proposition 8.2, we saw that every complete graph K_{n+1} has an ASD into paths. Many other graphs of size $\binom{n+1}{2}$ have this property as well. One other is the famous Petersen graph. In fact, the Petersen graph is a member of a class of

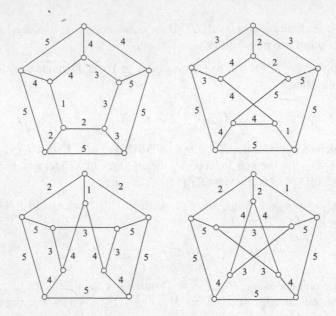

Fig. 8.5 The five permutation graphs of C_5 are path-perfect

four graphs called permutation graphs of C_5 (see [2]), all of which appeared on the cover of the original edition of the 1969 book *Graph Theory* by Harary [13]. Indeed, each of these graphs can be decomposed into five paths G_1, G_2, G_3, G_4, G_5 where $G_i = P_{i+1}$ has size i for $i = 1, 2, 3, 4, 5$, as illustrated in Fig. 8.5 where each edge in G_i is labeled i.

A graph G of size $\binom{n+1}{2}$ is called *path-perfect* if G can be decomposed into paths of lengths $1, 2, \ldots, n$. Consequently, every path-perfect graph of size $\binom{n+1}{2}$ has an ASD into n paths. With this terminology, every complete graph and every permutation graph of C_5 is path-perfect. It was shown by Fink and Straight [9] that certain complete bipartite graphs are also path-perfect.

Theorem 8.5 *For every positive integer r, the graphs $K_{r,2r+1}$ and $K_{r,2r-1}$ have an ASD into paths.*

For example, if $r = 3$, it follows by Theorem 8.5 that $K_{3,5}$ and $K_{3,7}$ are path-perfect. This is illustrated in Fig. 8.6, where an edge labeled i belongs to a path of length i in an ASD of $K_{3,5}$ and $K_{3,7}$, respectively.

While some graphs of size $\binom{n+1}{2}$, such as K_{n+1}, do not have an ASD into matchings, there are conditions under which a graph of size $\binom{n+1}{2}$ does have such an ASD. One such condition dealing with the chromatic index $\chi'(G)$ of a graph G was obtained by Chen [5].

Theorem 8.6 *If G is a graph of size $\binom{n+1}{2}$ where $n \geq 2\chi'(G) - 2$, then G has an ASD into matchings.*

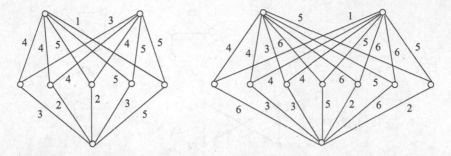

Fig. 8.6 The path-perfect graphs $K_{3,5}$ and $K_{3,7}$

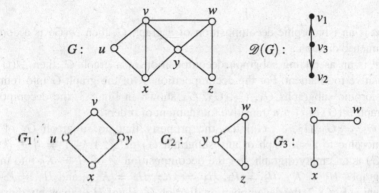

Fig. 8.7 A decomposition digraph

Without requiring an ASD into matchings, a lower bound for n was obtained in [7] in terms of the maximum degree $\Delta(G)$ of a graph G of size $\binom{n+1}{2}$.

Theorem 8.7 *If G is a graph of size $\binom{n+1}{2}$ where $n > \left(1 + \frac{1}{\sqrt{2}}\right)\Delta(G)$, then G has an ASD.*

8.4 Decomposition Digraphs

For a given graph G and a decomposition $\mathscr{D} = \{G_1, G_2, \ldots, G_k\}$ of G, there is a corresponding digraph $\mathscr{D}(G)$ with vertex set $V(\mathscr{D}(G)) = \{v_1, v_2, \ldots, v_k\}$ such that (v_i, v_j) is an arc of $\mathscr{D}(G)$ where $1 \le i \ne j \le k$ if G_i is isomorphic to a subgraph of G_j. The digraph $\mathscr{D}(G)$ is referred to as the *decomposition digraph of G* (with respect to decomposition $\mathscr{D}$ of G). For the decomposition $\mathscr{D} = \{G_1, G_2, G_3\}$ of the graph G of Fig. 8.7, the decomposition digraph $\mathscr{D}(G)$ of G has order 3 whose underlying graph is P_3, which is also shown in Fig. 8.7.

The following are three observations dealing with decomposition digraphs.

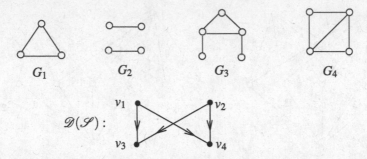

Fig. 8.8 The digraph $\mathscr{D}(\mathscr{S})$ of a set $\mathscr{S}$ of four graphs

- If $\mathscr{D}$ is an isomorphic decomposition of a graph G, then $\mathscr{D}(G)$ is a complete symmetric digraph.
- If $\mathscr{D}$ is an ascending subgraph decomposition of a graph G, then $\mathscr{D}(G)$ is a transitive tournament. For the decomposition $\mathscr{D}$ of the graph G into four non-isomorphic subgraphs G_1, G_2, G_3, G_4 shown in Fig. 8.3, the decomposition digraph $\mathscr{D}(G)$ of G is a transitive tournament of order 4.
- If $\mathscr{D} = \{G_1, G_2, \ldots, G_k\}$ has the property that no subgraph G_i of $\mathscr{D}$ is isomorphic to a subgraph of any subgraph G_j of $\mathscr{D}$, $1 \le i \ne j \le k$, then $\mathscr{D}(G)$ is an empty digraph. For the decomposition $\mathscr{D}$ of $H = K_6$ into the five subgraphs $H_1 = K_3$, $H_2 = P_4$, $H_3 = 3K_2$, $H_4 = K_{1,3}$, and $H_5 = P_2 + P_3$ shown in Fig. 8.2, the decomposition digraph $\mathscr{D}(H)$ of H is an empty digraph of order 5.

A digraph is referred to as a *decomposition digraph* if it is isomorphic to $\mathscr{D}(G)$ for some graph G with respect to a decomposition $\mathscr{D}$ of G. Among the questions involving decomposition digraphs are the following:

1. Which digraphs are decomposition digraphs?
2. For a decomposition digraph D, what is the smallest order of a graph G for which $\mathscr{D}(G) \cong D$?

There is a related concept not dealing with decompositions. For a set $\mathscr{S} = \{G_1, G_2, \ldots, G_k\}$ of graphs, there is an associated digraph $\mathscr{D}(\mathscr{S})$ called the *set digraph of $\mathscr{S}$* for which $V(\mathscr{D}(\mathscr{S})) = \{v_1, v_2, \ldots, v_k\}$ such that (v_i, v_j) is an arc of $\mathscr{D}(\mathscr{S})$, $i \ne j$, if G_i is isomorphic to a subgraph of G_j. For example, if $\mathscr{S} = \{G_1, G_2, G_3, G_4\}$ is the set of the four graphs shown in Fig. 8.8, then the digraph $\mathscr{D}(\mathscr{S})$ of $\mathscr{S}$ has order 4 and its underlying graph is C_4.

If $\mathscr{S} = \{G_1, G_2, G_3, G_4, G_5\}$ where G_1 is the graph of order 6 obtained by adding a pendant edge at each vertex of K_3, $G_2 = C_5 \vee K_1$, $G_3 = C_5$, $G_4 = P$ (the Petersen graph), and $G_5 = P_9$, then $\mathscr{D}(\mathscr{S})$ is the digraph whose underlying graph is $P_5 = (v_1, v_2, v_3, v_4, v_5)$ with four arcs (v_1, v_2), (v_3, v_2), (v_3, v_4), (v_5, v_4).

Among the problems dealing with set digraphs is the following:

3. For a given positive integer k, determine the smallest positive integer n for which there is a set $\mathscr{S}$ of k graphs, each of order at most n, such that $\mathscr{D}(\mathscr{S})$ is empty.

While there is clearly such an integer n for each positive integer k, there is no set $\mathscr{S}$ for which $\mathscr{D}(\mathscr{S})$ is an infinite empty digraph, which is a consequence of the following well-known theorem due to Robertson and Seymour [16]

The Graph Minor Theorem *In every infinite set of graphs, there are two graphs where one is (isomorphic to) a minor of the other.*

8.5 Monochromatic Ascending Subgraph Sequences

If G is a graph of size $m \geq 3$, then $\binom{n+1}{2} \leq m < \binom{n+2}{2}$ for a unique integer $n \geq 2$. A sequence $G_1, G_2, \ldots, G_k$ of k subgraphs of G where $k \leq n$ is an *ascending subgraph sequence* of G if

1. G_i has size i for $1 \leq i \leq k$,
2. every two subgraphs G_i and G_j, $1 \leq i \neq j \leq k$, are edge-disjoint, and
3. G_i is isomorphic to a subgraph of G_{i+1} for $1 \leq i \leq k-1$.

The *ascending subgraph index* $AS(G)$ of G is the maximum integer k for which G has an ascending subgraph sequence $G_1, G_2, \ldots, G_k$. The ASD conjecture states that $k = n$ for every graph G of size m where $m = \binom{n+1}{2}$.

A Ramsey theory problem related to these concepts was introduced in [3] and further studied in [4]. Let G be a graph of size m where $\binom{n+1}{2} \leq m < \binom{n+2}{2}$. The *ascending Ramsey index* $AR(G)$ of G is the maximum integer k such that for every red-blue coloring of the edges of G, there exists an ascending subgraph sequence $G_1, G_2, \ldots, G_k$ such that G_i is monochromatic for $1 \leq i \leq k$. A graph G of size $\binom{n+1}{2}$ is said to have a *monochromatic ascending subgraph decomposition* (or a *monochromatic ASD*) if for every red-blue coloring of the edges of G, there exists an ascending subgraph decomposition $G_1, G_2, \ldots, G_n$ of G such that each subgraph G_i is monochromatic for $1 \leq i \leq n$. Consequently, if a graph G of size $\binom{n+1}{2}$ has a monochromatic ASD, then $AR(G) = n$. This concept is illustrated in the next two examples.

Example 8.1 The graph K_4 has a monochromatic ASD and so $AR(K_4) = 3$.

Proof Let there be given an arbitrary red-blue coloring of K_4, resulting in the red subgraph G_R and the blue subgraph G_B of sizes m_R and m_B, respectively, where $m_R \leq m_B$. We show that K_4 has a monochromatic ASD. Since $m_R \leq m_B$, it follows that $0 \leq m_R \leq 3$. For $m_R \in \{0, 1\}$, such an ASD is clear. Suppose that $m_R = 2$. Then either $G_R = 2K_2$ or $G_R = P_3$. In either case, there is a monochromatic ASD with $G_1 = K_2$, $G_2 = G_R$, and $G_3 = P_4$. If $m_R = 3$, then $G_R \in \{K_3, K_{1,3}, P_4\}$. In each case, $G_1 = K_2$, $G_2 = K_{1,2}$, $G_3 = G_R$ is a monochromatic ASD of K_4. $\square$

Example 8.2 The graph $G = 3K_2 + K_{1,7}$ of size 10 has ascending Ramsey index 3.

Fig. 8.9 The
graph $G = 3K_2 + K_{1,7}$

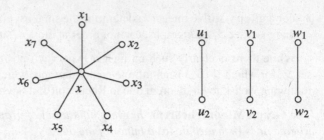

Proof First, we show that $AR(G) \neq 4$. Label the vertices of G as shown in Fig. 8.9. Consider the red-blue coloring of G, resulting in the red subgraph $G_R \cong 4K_2$ where $E(G_R) = \{u_1u_2, v_1v_2, w_1w_2, xx_1\}$. We show that there is no monochromatic ASD of G into four graphs G_1, G_2, G_3, G_4 of G with this red-blue coloring, for suppose that there is. Then either (1) only G_4 is a red subgraph or (2) only G_1 and G_3 are red subgraphs. We consider these two cases.

Case 1 Only G_4 is a red subgraph of G. Since $G_4 = 4K_2$, it follows that $G_3 = K_{1,3}$. Because $K_{1,3}$ is not isomorphic to a subgraph of G_4, this is a contradiction.

Case 2 Only G_1 and G_3 are red subgraphs of G. Since $G_1 = K_2$ and $G_3 = 3K_2$, it follows that $G_2 = K_{1,2}$ and $G_4 = K_{1,4}$. Because G_3 is not isomorphic to a subgraph of G_4, for example, this is a contradiction.

Therefore, $AR(G) \leq 3$. It remains to show that $AR(G) \geq 3$. Let there be given an arbitrary red-blue coloring of G. Let $G_1 = K_2$ be any of the three components of size 1 in G and let $G_2 = K_{1,2}$ be a monochromatic subgraph of $K_{1,7}$. The remaining subgraph $K_{1,5}$ of $K_{1,7}$ has three edges colored the same. Let $G_3 = K_{1,3}$ be such a monochromatic subgraph of $K_{1,5}$. Then G_1, G_2, G_3 is a monochromatic ascending subgraph sequence in G. Thus, $AR(G) \geq 3$ and so $AR(G) = 3$.   □

If G is a star or a matching of size $\binom{n+1}{2}$, then G has a monochromatic ASD and consequently, $AR(G) = n$, as we show next.

Proposition 8.4 *For each integer $n \geq 2$, every star of size $\binom{n+1}{2}$ has a monochromatic ASD.*

Proof We proceed by induction on n. The truth of this statement is immediate for $n = 2$. Assume for an arbitrary integer $n \geq 2$ that every star of size $\binom{n+1}{2}$ has a monochromatic ASD. Let G be a star of size $\binom{n+2}{2}$ and let there be given a red-blue coloring of G. Since $n \geq 2$, it follows that $\frac{1}{2}\binom{n+2}{2} \geq n + 1$ and so there is a monochromatic substar H of G having size $n + 1$. Let U be the set of the $n + 1$ end-vertices of the substar H and let $G' = G - U$. Thus, G' is a star of size $\binom{n+1}{2}$. By the induction hypothesis, the resulting red-blue coloring of G' has a monochromatic ASD into n monochromatic subgraphs $G_1, G_2, \ldots, G_n$. Hence, $G_1, G_2, \ldots, G_n$, $G_{n+1} = H$ is a monochromatic ASD of G.   □

The following result has a proof similar to that of Proposition 8.4.

Proposition 8.5 *For each integer* $n \geq 2$, *every matching of size* $\binom{n+1}{2}$ *has a monochromatic ASD.*

Among the numerous problems on this topic are the following.

1. In addition to stars and matchings of size $\binom{n+1}{2}$, which graphs G of size $\binom{n+1}{2}$ have $AR(G) = n$?
2. For which pairs k, n of positive integers with $k \leq n$, does there exist a graph G of size $\binom{n+1}{2}$ such that $AR(G) = k$?

References

1. Y. Alavi, A. J. Boals, G. Chartrand, P. Erdős, O.R. Oellermann, The ascending subgraph decomposition problem. Congr. Numer. **58**, 7–14 (1987)
2. G. Chartrand, F. Harary, Planar permutation graphs. Ann. Inst. H. Poincaré (Sect. B) **3**, 433–438 (1967)
3. G. Chartrand, P. Zhang, Ramsey sequences of graphs. AKCE J. Graphs Combin. **17**, 646–652 (2020)
4. G. Chartrand, P. Zhang, New directions in Ramsey theory. Discrete Math. Lett. **6**, 84–96 (2021)
5. H. Chen, On the ascending subgraph decomposition into matchings. J. Math. Res. Expos. **14**(1), 61–64 (1994)
6. H. Chen, K. Ma, On the ascending subgraph decompositions of regular graphs. Appl. Math. J. Chin. Univ. **13B**, 165–170 (1998)
7. R.J. Faudree, R. Gould, M.S. Jacobson, L. Lesniak, Graphs with an ascending subgraph decomposition. Congr. Numer. **65**, 33–42 (1988)
8. R.J. Faudree, A. Gyárfás, R.H. Schelp, Graphs which have an ascending subgraph decomposition. Congr. Numer. **59**, 49–54 (1987)
9. J.F. Fink, H.J. Straight, A note on path-perfect graphs. Discrete Math. **33**, 95–98 (1981)
10. H. Fu, Some results on the ascending subgraph decomposition. Bull. Inst. Math. Acad. Sin. **16**(4), 341–345 (1988)
11. H. Fu, A note on the ascending subgraph decomposition problem. Discrete Math. **84**, 315–318 (1990)
12. J.A. Gallian, A dynamic survey of graph labeling. Electron. J. Combin. (2017), #DS6.
13. F. Harary, *Graph Theory* (Addison-Wesley, Reading, 1969)
14. K. Ma, H. Zhou, J. Zhou, On the ascending star subgraph decomposition of star forest. Combinatorica **14**(3), 307–320 (1994)
15. G. Ringel, Problem 25, in *Theory of Graphs and Its Applications* (Nakl. ČSAV, Prague, 1964), p. 162
16. N. Robertson and P. D. Seymour, Graph minors. XX Wagner's conjecture. J. Combin. Theory Ser. B **92**, 325–357 (2004)
17. A. Rosa, On certain valuations of the vertices of a graph, in *Theory of Graphs* (Gordon and Breach, New York, 1967), pp. 349–355
18. R.M. Wilson, Decompositions of complete graphs into subgraphs isomorphic to a given graph. Congr. Numer. **15**, 647–659 (1976)

Index

Printed in the United States
by Baker & Taylor Publisher Services